NELSON
English 7

NELSON

NELSON

COPYRIGHT © 2019 by Nelson Education Ltd.

ISBN-13: 978-0-17-686953-3
ISBN-10: 0-17-686953-0

Printed and bound in Canada
1 2 3 4   22 21 20 19

For more information, contact Nelson Education Ltd., 1120 Birchmount Road, Toronto, Ontario M1K 5G4. Or you can visit our website at nelson.com.

ALL RIGHTS RESERVED. No part of this publication may be reproduced, stored in a retrieval system, or transmitted in any form or by any means, electronic, mechanical, photocopying, scanning, recording or otherwise, except as specifically authorized.

For permission to use material from this text or product, submit all requests online at cengage.com/permissions. Further questions about permissions can be emailed to permissionrequest@cengage.com.

Every effort has been made to trace ownership of all copyrighted material and to secure permission from copyright holders. In the event of any question arising as to the use of any material, we will be pleased to make the necessary corrections in future printings.

Cover image: Blend Images – JGI / Jamie Grill / Brand X Pictures / Getty Images

TABLE OF CONTENTS

Work with Vocabulary 5
1. Use a Similar Word: **Synonyms** 6
2. Use the Opposite Word: **Antonyms** 7
3. Choose the Correct Spelling:
   Homophones 8
4. Expand Your Vocabulary: **Root Words** 10
5. Understand Word Beginnings: **Prefixes** 11
6. Understand Word Endings: **Suffixes** 12
7. Combine Two Words: **Contractions** 13
8. Mind Your Meaning:
   Denotation and Connotation 14
9. Use Strong Words: **Nouns and Verbs** 15
10. Use Colloquialisms:
    Informal and Formal Language 17
11. Use Variety: **Literary Devices** 19
Section Review 21

Build Sentences 25
12. Use Variety: **Types of Sentences** 26
13. Use Variety: **Sentence Length** 28
14. Combine Sentences:
    Compound Sentences 29
15. Expand Sentences: **Adding Details** 31
16. Edit Sentences: **Run-On Sentences** 33
17. Know **Complete Subjects**
    and Predicates 34
18. Identify Who or What: **Simple Subjects** 36
19. Identify the Action: **Simple Predicates** 37
20. Identify Sentence Parts:
    Direct and Indirect Objects 38
21. Recognize **Independent**
    and Subordinate Clauses 39
22. Combine Sentences:
    Complex Sentences 40
23. Recognize Clauses: **Adjective Clauses** 42
24. Recognize Clauses: **Adverb Clauses** 43
25. Edit Sentences: **Sentence Fragments** 44
26. Edit Sentences: **Comma Splices** 45
Section Review 46

Know Capitalization and Punctuation 50
27. Use Capitals:
    A Variety of Capitalization 51
28. Use a Dictionary: **Abbreviations** 53
29. Identify Short Forms: **Abbreviations** 54
30. Use Variety: **Commas** 55
31. Punctuate Dialogue: **Quotation Marks** 57
32. Show Possession: **Apostrophes** 59
33. Join Independent Clauses: **Semicolons** 60
34. Separate Titles and Subtitles: **Colons** 61
35. Add Less Important Information:
    Parentheses 62
36. Guide Readers:
    A Variety of Punctuation 63
Section Review 65

Grasp Grammar and Usage 69
37. Name the Person, Place,
    Thing, or Idea: **Nouns** 70
38. Show Ownership:
    Singular Possessive Nouns 72
39. Show Group Ownership:
    Plural Possessive Nouns 73
40. Use **Irregular Plural**
    Possessive Nouns 74
41. Use **Concrete and**
    Abstract Nouns 75
42. Identify **Action, Auxiliary,**
    and Linking Verbs 76
43. Provide More Information:
    Verb Phrases 78
44. Show When an Action Happens:
    Verb Tenses 79
45. Make the Past Tense:
    Irregular Verbs 81
46. Use **Present Perfect and**
    Past Perfect Tenses 82
47. Match the Numbers:
    Subject–Verb Agreement 83

48. Match the Subject: **Linking Verbs** 85
49. Replace Subject Nouns:
    Subject Pronouns 86
50. Replace Object Nouns:
    Object Pronouns 87
51. Show Ownership:
    Possessive Pronouns 88
52. Use **Indefinite Pronouns** 89
53. Use **Reflexive Pronouns** 90
54. Make **Pronouns
    and Antecedents Agree** 92
55. A Variety of **Pronouns
    and Antecedents** 93
56. Write Descriptive Words: **Adjectives** 94
57. Make Comparisons: **Adjectives** 96
58. Describe Actions: **Adverbs** 97
59. Make Comparisons: **Adverbs** 98
60. Write Descriptively:
    Adjectives and Adverbs 99
61. Show Relationships: **Prepositions** 100
62. Recognize Phrases:
    Prepositional Phrases 101
63. Describe a Noun: **Participle Phrases** 102
64. Be Clear: **Misplaced Modifiers** 103
65. Be Clear:
    Misplaced and Dangling Modifiers 104
66. Use Joining Words: **Conjunctions** 105
67. Express Emotions: **Interjections** 106
Section Review 107

Craft and Compose 110
68. Create a Life Map: **Choosing a Topic** 111
69. Choose Your Voice:
    Purpose and Audience 112
70. State Your Purpose:
    Topic and Thesis 113
71. Cluster with a Web: **Organizing Ideas** 114
72. Use a Graphic Organizer:
    Organizing Ideas 115
73. Use Dialogue: **Strong Openings** 117
74. Lead with a Statistic: **Strong Openings** 118
75. Use Your Senses: **Writing Details** 119
76. Use Examples: **Writing Details** 121

77. Use Variety: **Writing Details** 123
78. Format a Speaker's Words:
    Writing Dialogue 125
79. Make Language Precise:
    Avoiding Redundancies 127
80. Sum Up Your Narrative:
    Strong Conclusions 128
81. Sum Up Your Report:
    Strong Conclusions 129
82. Catch Your Readers' Attention:
    Effective Titles 131
83. Check for Errors:
    Revising and Editing 132
84. Correct Sentences:
    Revising and Editing 134
Section Review 136

Develop Research Skills 139
85. Have a Clear Focus:
    Inquiry Questions 140
86. Find Synonyms:
    Researching Words 141
87. Choose Resources:
    Library Research 142
88. Conduct Online Research:
    Keywords 143
89. Know the Difference:
    Primary and Secondary Sources 144
90. Look for Consensus:
    Evaluating Websites 145
91. Be Critical:
    Evaluating Websites 146
92. Remix and Rework:
    Plagiarism and Copyright 147
93. Citations for Videos: **Crediting Sources** 148
94. Track Print and Online Sources:
    Research Notes 149
95. Use Ideas and Words:
    Paraphrasing and Quoting 151
Section Review 153

Answer Key 157
Graphic Organizers 172
Index 176

WORK WITH VOCABULARY

Words are the basic units of meaning. When we learn how to use words, we are learning how to describe the world around us.

If you pay attention to the meaning each word can have, you will notice that most words can mean different things, depending on how they are used.

Taking one word and combining it with others, or adding or removing parts, can have interesting effects. For example, consider the word *scene*. Now think about the meanings of *scenic, obscene, scenery,* and *causing a scene*.

When good writers choose their words carefully, they give their readers a clear picture or feeling.

In this section, you will learn how to use words to write exactly what you mean.

> "Words are important. If you cannot say what you mean, you will never mean what you say. And you should always mean what you say."
> — Mark Peploe and Bernardo Bertolucci

USE A SIMILAR WORD: SYNONYMS

If you want your readers to remain interested in your writing, make sure you do not repeat the same words too often. Try using **synonyms** instead. A synonym is a word with the same, or almost the same, meaning as another word.

For example: Reese was supposed to *lead* the campers to the lake.
Instead of ~~leading~~ *guiding* them, however, she
gave them a map and sent them into the woods.

Note that the context in which a word is used may narrow your synonym choices. Even though "leading" can also mean "winning," as in a sport, you would not write, "Instead of ~~leading~~ *winning* them ..."

A. For each underlined word, circle the word underneath that is closest in meaning. Pay attention to the context.

1. <u>complete</u> darkness
   a) incomplete
   b) absolute *(circled)*
   c) finished

2. <u>accept</u> the new member
   a) refuse
   b) agree *(circled)*
   c) welcome

3. <u>expand</u> your knowledge
   a) increase *(circled)*
   b) decrease
   c) exchange

B. For each sentence, two synonyms are provided for the underlined word. Circle the synonym that best fits the context.

1. No building will stand for long without a solid <u>foundation</u>. (**base** / basis)

2. Weather experts <u>categorize</u> each hurricane according to its wind intensity. (arrange / **classify**)

3. The general's health declined as the <u>stress</u> of the war continued to take its toll. (emphasis / **pressure**)

C. Rewrite each sentence, replacing the underlined word with a synonym. You may need to use a thesaurus to help you complete this exercise.

1. We need to find a <u>site</u> for our next game.

   _____

2. Mayor Okoro could sense <u>unrest</u> in the crowd.

   _____

D. Go back to a previous piece of your writing. Choose three nouns and three verbs and think of several synonyms for each one (or use a thesaurus to look them up). Experiment with replacing the words in your piece of writing with different synonyms, and then, choose the best synonym for each context.

6   Work with Vocabulary   Copyright © 2019 by Nelson Education Ltd.

LESSON 2

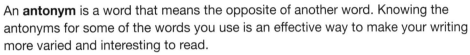
USE THE OPPOSITE WORD: ANTONYMS

An **antonym** is a word that means the opposite of another word. Knowing the antonyms for some of the words you use is an effective way to make your writing more varied and interesting to read.

For example: Nina was not *happy* with her team's *laziness*.
In fact, she was quite *disappointed* by their lack of *energy*.

In the example, *happy* and *disappointed* are antonyms, and so are *laziness* and *energy*.

A. Complete each sentence by choosing an antonym for the word in parentheses ( ).
Use the words in the following list:

~~unusual~~   ~~investigate~~   ~~foolish~~

1. I knew it was (wise) __foolish__ for Wes to play hockey without his helmet.
2. His behaviour had been (normal) __unusual__ all day.
3. I decided to talk to Wes and (ignore) __investigate__ what might be troubling him.

B. In the following paragraph, cross out each underlined term and write an antonym above it.

I have <u>never</u> liked comedies. I <u>loathe</u> them! In my opinion, they are the <u>worst</u> movie genre. Some of my friends really <u>enjoy</u> watching comedies, though. They <u>never</u> watch action movies or dramas. They think they are <u>worse</u>. I think they are <u>right</u>!

C. Write a sentence using both of the antonyms listed below.

1. anxious, calm

___
___

2. refuse, agree

___
___

D. Write a paragraph or two about your favourite food. Think about the words you have used that have antonyms. How can you use those antonyms effectively to make your paragraph more varied and interesting to read? Revise your work.

LESSON 3
CHOOSE THE CORRECT SPELLING: HOMOPHONES

Homophones are words that sound exactly like one another but have different meanings and sometimes different spellings. When you write, it is important to use the word with the appropriate meaning and spelling. You can confuse your reader if you write the wrong homophone!

For example: The counsellor read the campers' cabin assignments *aloud*. ✓
The counsellor read the campers' cabin assignments *allowed*. ✗

A. Read each sentence and decide if the underlined homophone is spelled correctly. If it is correct, underline *Yes*. If it is not correct, underline *No* and write the correct spelling on the line provided.

1. In Europe during the 1600s, it was very fashionable to <u>ware</u> fur.   Yes   No _____

2. Fur traders in Canada struggled <u>too</u> keep up with the demand.   Yes   No _____

3. Beaver furs could be sold for quite a large <u>sum</u> of money.   Yes   No _____

4. By 1800, however, the fur trade in Canada was becoming <u>week</u>.   Yes   No _____

B. Complete each sentence with a pair of homophones. Write one homophone on each line in the sentence. Write them in the correct place. Choose from the following pairs:

their / there   pour / pore   stationery / stationary   heard / herd

1. As the rain began to _____ down, Thompson sat at the table to _____ over his map.

2. Since he was _____ until the rain stopped, he pulled out a sheet of _____ and began a letter.

3. He wrote, "How is everyone _____? Do Elizabeth and Henry miss _____ father?"

4. "Last week, I told you about a thunderous noise I _____ during the night, which was in fact a _____ of caribou!"

C. Write a sentence using each set of homophones provided.

1. your, you're _____
_____

2. its, it's _____
_____

3. too, to _____
_____

8   Work with Vocabulary

D. Write two descriptive paragraphs about living in your community. Include as many homophones as you can. To help you get started, consider using some of the following homophones: its / it's; their / there / they're; write / right; new / knew.

EXPAND YOUR VOCABULARY: ROOT WORDS

Words with many syllables and parts can be difficult to understand and spell. To expand the number of words you can understand and spell, become familiar with some commonly used **root words**. Like base words, most root words are word parts that may or may not stand alone as words. For this reason, you usually add a prefix, suffix, or both to a root word.

For example: The word *postscript* has the root word *script*, which means "to write." It also has the prefix *post-*, which means "after." So, a *postscript* is <u>written after</u> something else.

Many root words come from other languages, such as the following:
script and *scribe* ("to write") come from Latin
spec / *spect* ("to look") and *vis* / *vid* ("to see") come from Latin
tele ("far off") and *gen* ("birth") come from Greek

A. Identify the common root word of each group of words. Write the root word and its meaning on the lines provided.

	Root Word	Meaning of Root Word
1. envision, vista, visual	_____	_____
2. prospect, speculate, inspection	_____	_____
3. describe, scribble, prescribe	_____	_____

B. Read each sentence and decide if it is true or false, and briefly explain your decision. Use the root word of the underlined word to help you.

1. <u>Genealogy</u> is the study of eyesight. _____
   _____

2. If something is <u>evident</u>, it is easy to see. _____
   _____

3. To view distant objects, you might want to use a <u>telescope</u>. _____
   _____

C. Write three more words that contain each root word listed below.

1. *gen*: _____
2. *tele*: _____

D. Find a root word not covered in this lesson and list five words that contain it.

LESSON 5
UNDERSTAND WORD BEGINNINGS: PREFIXES

A **prefix** is a word part that comes at the beginning of a base word.
For example: *interchange* = the prefix *inter-* + the base word *change*

When you add a prefix before a base word, the meaning of that word changes.
For example: The prefix *inter-* means *between* or *among*. The word *interchange* can mean "to *change* places of or between two things."

Knowing the meaning of common prefixes can help you understand and write unfamiliar words.
The prefix *auto-* means "self."
The prefix *anti-* means "against" or "opposed to."
The prefix *pro-* means "before / in front of" or "in support of."

A. **For each underlined word, circle the correct definition.**

1. An <u>antibacterial</u> cream works …
    a) in support of bacteria.
    b) against bacteria.
    c) to move bacteria forward.

2. An <u>autobiography</u> is …
    a) a biography written by a stranger.
    b) a biography about automobiles.
    c) a biography written by the subject of the book.

3. A book's <u>prologue</u> …
    a) contains comments against the book.
    b) comes between two chapters.
    c) comes before Chapter 1.

B. **Complete the sentences by adding a prefix to the word in parentheses and writing the new word on the line provided. Choose from these prefixes:**

   inter-   auto-   anti-

1. After weeks of ignoring one another, the two dogs were finally _____ (acting).
2. A large crowd attended the _____ (war) rally and carried signs.
3. The company _____ (viewed) ten people for the job.
4. Eli was thrilled to get his favourite football player's _____ (graph).

Work with Vocabulary 11

LESSON 6
UNDERSTAND WORD ENDINGS: SUFFIXES

A **suffix** is a word part that is added to the end of a base word. Sometimes, you need to change the spelling of the base word when you add a suffix.

For example: *conclusion* = the base word *conclude* − the last two letters *de* + the suffix *-sion*

Adding a suffix to the end of a base word can change the meaning of that word. Knowing the meaning of common suffixes can help you understand and spell unfamiliar words.

For example: The suffix *-sion* means "a state or quality of." If something has reached its *conclusion*, it has the quality of being *concluded*.

The suffixes *-ence* and *-ance* can also mean "a state or quality of" or "the action of." The suffix *-graph* means "something that writes or records" or "the record of."

A. Complete the word equations. On each line provided, write a base word and/or a suffix to make a new word. Choose from these suffixes:

-sion   -ence   -ance   -graph

1. explode − _____ + _____ = explosion
2. insure − _____ + -ance = _____
3. _____ + -ence = insistence
4. persist + -ence = _____
5. auto + _____ = autograph
6. erode − _____ + -sion = erosion

B. Using the suffixes in the words on the left as clues, draw a line between each word and the correct definition on the right.

division   the written study of a single subject

adolescence   the state of being a teenager

monograph   the action of separating into parts

C. Brainstorm three words that have the suffix *-ance* and write them on the lines below. Then, write a sentence using each word.

_____   _____   _____

_____

_____

_____

D. As you read, make a list of words that have the suffix *-sion*, *-ence*, *-ance*, or *-graph*. Write definitions for the words you find.

LESSON 7: COMBINE TWO WORDS: CONTRACTIONS

Using **contractions** when you speak makes your speech sound natural. Likewise, using contractions when you write makes the tone of your writing more informal.

To form a contraction when you write, combine two or more words to make a new word. Then, replace one or more of the letters with an apostrophe (') to make the new word shorter.

Sometimes, different combinations of words can form the same contraction. *He had* and *he would* both form *he'd* as a contraction.

For example: *He'd* (he had) left Ireland in 1862. *He'd* (he would) never go back!

Sometimes, you need to use context clues to understand which two words form the contraction. An *'s* can indicate *has*, *us*, or *is*.

For example: *He's* (he has) immigrated to Canada.
Let's (let us) go, too! *It's* (it is) an exciting opportunity.

A. **Complete each word equation.**

1. I + had = _____
2. _____ + would = we'd
3. that + _____ = that's
4. _____ + would = I'd
5. we + had = _____
6. it + is = _____

B. **After each contraction, write the words that have been combined.**

1. They'd (_____) arrived by ship two days earlier than expected.
2. They'd (_____) remain in Canada for the rest of their lives.
3. She'd (_____) always tell us stories about England before we went to sleep.
4. She'd (_____) so many interesting things to teach us about life.

C. **Rewrite each sentence, replacing the underlined words with a contraction.**

1. <u>There is</u> a plaque commemorating the Canadian Pacific Railway.

_____

_____

2. <u>It has</u> been more than 125 years since completion of the railway.

_____

_____

LESSON 8
MIND YOUR MEANING: DENOTATION AND CONNOTATION

Good writers choose their words carefully in order to convey a clear meaning. Many words have both a **denotation** and a **connotation**. The denotation of a word is its exact dictionary meaning. The connotation of a word is the emotional feeling that a word holds. These feelings can be positive or negative.

Two words may have similar denotations, but different connotations. Writers need to be sure that their words have the intended effect on readers' feelings.

For example: My *curious* brother asks a lot of questions.
My *nosy* brother asks a lot of questions.

In these examples, the synonyms *curious* and *nosy* have the same denotation; they both basically mean "inquisitive." However, *curious* has a positive connotation, while *nosy* has a negative one.

A. **To complete each sentence, choose the noun with the most *positive* connotation from the list underneath. Underline your choice.**

1. Deleting that important file was a/an—
    a) error
    b) misunderstanding
    c) mistake
    d) blunder

2. I could hear the neighbours having a/an—
    a) debate
    b) argument
    c) disagreement
    d) fight

B. **Rewrite the sentences in two different ways. First, replace the underlined word with a synonym that has a *positive* connotation. Then, replace it with a synonym that has a *negative* connotation.**

1. Wow, your puppy is <u>active</u>!

    Positive: _____

    Negative: _____

2. That's an <u>interesting</u> hat.

    Positive: _____

    Negative: _____

C. **Go back to a piece of your writing. Experiment with some of the nouns you used by replacing them with words having a different connotation. Write the new sentences and indicate whether each has a more positive or negative connotation because of the synonym.**

14   Work with Vocabulary

USE STRONG WORDS: NOUNS AND VERBS

Writers want their work to be clear and interesting. So do their readers! Here are some ways to use **strong nouns and verbs** to liven up your writing:

- Use *specific* nouns and verbs, not general ones.
  For example: The ~~police~~ North West Mounted Police enforced the law in Canada's West.
- Use *vivid* nouns and verbs to produce strong feelings or clear images.
  For example: Several officers ~~went~~ travelled to Yukon to control the ~~group~~ stampede of gold seekers.
- Change *wordy* verbs that are followed by nouns ending in *-tion, -ance, -ence,* or *-ment* into more concise verbs.
  For example: They ~~gave reinforcement to~~ reinforced the officers already stationed there.
- Replace a *group* of words with a single, more effective, noun or verb.
  For example: As a result, the Gold Rush did not cause ~~a lot of disruption~~ chaos.

A. **Read each sentence and underline the word in each pair that is more specific or more vivid.**

1. By the 1850s, the inhabitants of the (settlement / Red River Colony) were mostly Métis.

2. Led by Louis Riel, they fought to (preserve / have) their homes, culture, and religion.

3. In an attempt to resolve the (issue / hostilities), the Canadian government and Riel's government began negotiating.

B. **For each group of words on the left, find a single word on the right that is more effective.**

1. speaking two languages   _____   contribute

2. make a contribution   _____   informed

3. provided information   _____   bilingual

C. **Rewrite each sentence, replacing the underlined group of words with a more effective noun or verb.**

1. People settled near the <u>body of water</u>.

_____

2. Women's groups across Canada <u>presented petitions</u>.

_____

D. Write a few persuasive paragraphs about whether Canada should or should not become part of the United States. Include as many strong nouns and verbs as you can.

LESSON 10: USE COLLOQUIALISMS: INFORMAL AND FORMAL LANGUAGE

Whenever you write, you should always know your audience and your purpose. This means knowing for whom you are writing and why you are writing. This knowledge will help you decide how formal your language should be.

If you are writing a history essay for school, you should use **formal language**. If you are writing an email to a friend, you could use **informal language**. Informal language is closer to the way we speak than formal language is. It can include simpler, shorter words, contractions, and colloquialisms.

Colloquialisms are expressions in which the words should not be taken literally.

For example: Informal language—I have a pretty good feeling that this car is *a hunk of junk*.

Formal language—I am almost certain that this car is not in working condition.

A. Underline each colloquialism below and write its meaning on the line beside it.

What's up? _____

I wasn't born yesterday. _____

He seems to be feeling blue. _____

"Can I borrow a toonie, please?" _____

B. Rewrite each sentence, replacing the underlined colloquialism with more formal language.

1. I'd like to return my salad. The hair in it is <u>grossing me out</u>.

_____

_____

2. Could you turn your music down? It is <u>driving me up the wall</u>.

_____

_____

C. In the paragraph below, underline each colloquialism. Above each one, replace the colloquialism with more formal language.

Yesterday, I saw a whole bunch of people gathered in the park. When I got closer, I saw that they had their eyes on a dude performing magic tricks. He was OK, but it was a no-brainer to see through most of what he was doing. When the guy passed around his hat and asked for loot at the end of his act, I took off.

D. Write two versions of a short conversation between two people arranging to meet later. In the first version, use informal language and include as many colloquialisms as you can. Then, rewrite the conversation using more formal language and no colloquialisms.

LESSON 11: USE VARIETY: LITERARY DEVICES

Writers use many tools to make their work come alive. **Literary devices** are tools that can help readers visualize what is being described so that they understand it better. Literary devices can also cause readers to experience a specific emotion or reaction. There are many types of literary devices.

For example: *Simile*—a comparison using the words *like* or *as*: *The child ate like a bird.*
Metaphor—a comparison not using the words like or as: *I ran through a sheet of rain.*
Idiom—an expression that means something different from its literal meaning: *That offer seems a little fishy.*
Alliteration—several words with the same beginning consonant sound appearing close together: *Tricky triangles tormented Trina.*
Personification—the use of human qualities or abilities to describe something non-human: *Opportunity knocks.*
Onomatopoeia—a word that sounds like what it means or that imitates the sound made by a person, animal, or thing: *boing, gurgle.*

A. **The sentences below all use literary devices. On the line beside each sentence, write whether the literary device is a simile, a metaphor, an idiom, alliteration, personification, or onomatopoeia.**

1. I climbed up to the high-dive platform, and then I chickened out. _____
2. The engine complained loudly as Mick sped to the hospital. _____
3. The snow fell softly and silently on city streets. _____
4. The flag whapped and flapped in the wind. _____
5. Her head was an encyclopedia of facts about early Canada. _____
6. Being on the bus was as boring as watching paint dry. _____

B. **Each line in this poem uses a literary device such as simile, personification, onomatopoeia, or alliteration. Underline the example of a literary device in each line and write the type of literary device on the line provided.**

Like a mother gently waking her babies, _____

the Sun peeks over the horizon _____

and nudges the sleeping flowers. _____

Stretching and straightening,
they soon stand tall. _____

C. Choose your two favourite types of literary devices. Write two sentences using them.

1. _____

2. _____

D. Write a poem about a difficult journey. Use literary devices such as similes, metaphors, idioms, alliteration, personification, or onomatopoeia.

_____
_____
_____
_____
_____
_____
_____
_____
_____
_____
_____
_____
_____
_____
_____
_____
_____
_____
_____
_____
_____
_____

SECTION REVIEW

A. Write each of the words below in the correct space on the chart:

whole   knowledge   shrink   ignorance
grow   partial   welcome   refuse

Word	complete	accept	expand	awareness
Synonym				
Antonym				

B. Complete each sentence with one of the following pairs of homophones. Use the correct homophone on each blank line.

seller / cellar   you're / your   pale / pail   its / it's   there / their

1. Even the _____ of the house was not aware of its secret _____.

2. Marla set down the _____ of _____ paint and chose a brush.

3. Does _____ neighbour know that _____ going out of town?

4. I think _____ funny when the dog chases _____ tail.

5. I put _____ dessert on the counter over _____.

C. Using a dictionary, look up the meaning of the following words with the root *tele-* (meaning "far off"). On the lines provided, explain how the meaning of *tele-* makes sense in each word.

  1. telemarketing   2. teleconference   3. telecommuting

1. _____
2. _____
3. _____

D. Find two additional words beginning with the root *tele-* and write them on the lines below. Then, explain how the root word gives you a clue to the meaning of each word.

tele- _____

tele- _____

E. Rewrite the sentence below. Instead of each contraction, write the two words that were combined to form the contraction.

That's the house we'd have bought—let's stop and see if it's changed at all.

_____

_____

Work with Vocabulary

F. For each sentence, choose the word in parentheses with the more positive connotation. Underline your choice.

1. Suki sized up her (*opponent, enemy*) and suddenly became very anxious.
2. We aim to give all of our (*customers, guests*) a pleasant dining experience.
3. Should all political parties have the right to distribute their (*information, propaganda*) on the streets?

G. Rewrite each sentence below, replacing the underlined groups of words with more effective nouns or verbs. Use the clues at the end of each sentence to help you.

1. Sabine told us the story of how <u>one of her relatives</u> survived a hurricane. (Use more specific words.)

   _____

   _____

2. The wind <u>blew</u> the roof off the house, and the rain <u>came</u> in. (Use more vivid words.)

   _____

   _____

3. Rescuers soon <u>came to the realization</u> that they would need helicopters. (Use strong verbs.)

   _____

   _____

4. All the <u>people who lived in the neighbourhood</u> cooperated. (Use a single word.)

   _____

   _____

H. Match each colloquialism on the left with its more formal version on the right. Draw a line to connect the two phrases.

to drive someone up the wall	to be very happy
to put your money where your mouth is	to joke with someone
to pull someone's leg	to irritate or frustrate someone
to be tickled pink	to act in a way that supports your opinion

I. Each line of the following piece of writing contains an idiom, a metaphor, and/or an example of personification. Write which is used beside each line.

A mountain of homework stares at me, _____

And a pit of fear opens in my stomach. _____

Am I in over my head? _____

"Keep your chin up," I tell myself. _____

"And use some elbow grease!" _____

Work with Vocabulary

J. Write your own non-rhyming poem about a very hot or a very cold day. Include literary devices such as similes, alliteration, and onomatopoeia.

K. **Choose one of the metaphors below, or a metaphor that you particularly like, and write an explanation of why it is true or false. Use strong nouns and verbs as often as possible. When you have finished writing, circle the strong nouns and verbs.**

Time is money.   Ideas are wings.   All the world's a stage.

BUILD SENTENCES

Sentences are where *things* and *actions* meet to make a complete idea. Each sentence is a tiny story in itself—nouns (or subjects) are the characters, and verbs (or predicates) are the plots.

In good writing, every sentence should have a reason for being there. Every sentence should have a purpose, whether it is to describe a scene, explain details, make a point, or make the reader feel an emotion.

In this section, you will learn how to put together perfect sentences.

> "A word after a word after a word is power."
> — Margaret Atwood

LESSON 12: USE VARIETY: TYPES OF SENTENCES

There are four **types of sentences**: declarative (makes a statement), imperative (makes a command), interrogative (asks a question), and exclamatory (makes an exclamation).

For example: We drove to Calgary in one hour. (statement)
Drive me to Calgary. (command)
How long did it take you to drive to Calgary? (question)
I want to go to Calgary! (exclamation)

Use different sentence types for different purposes and audiences. For example, commands are effective if you are writing instructions. Questions are often used in dialogue or to directly address a reader. Exclamations express strong emotion.

Each type of sentence requires specific punctuation. Statements and commands end with a period (although commands can sometimes end with an exclamation mark). Questions end with a question mark, and exclamations end with an exclamation mark.

A. Add the end punctuation to each sentence. Write the sentence type (declarative, imperative, interrogative, exclamatory) on the line.

1. We just put solar panels on our house___   _____
2. Stir the mixture until it is soft and creamy___   _____
3. I can't believe you fell for it ___   _____
4. Ask your mom if I can come over tonight___   _____
5. Who is coming to the museum with us tomorrow___   _____
6. Will you be able to help me with my science project tonight___   _____
7. My dad and I planted vegetables on the balcony___   _____
8. I'm so excited about the class trip___   _____
9. Where are you going to school next year___   _____
10. This soup tastes funny___   _____

B. Add the missing end punctuation to each sentence.

Last weekend, my dad took me and my friends camping___ It was the best weekend ever___ I had never been camping before, and I was worried that it would be boring___ But, it was so much fun___ The first night, Dad scared us all___ We were sitting around the campfire, and everything was quiet___ Then, he said, "Did anyone hear that___" We shook our heads___ "I swear, I just heard something," he said___ "Look over there___ Do you see anything___" Then, as we looked around, he snuck up on us and yelled, "Watch out___" We all screamed___ He scared us half to death___

C. Write eight sentences about the environment. Write two declarative sentences, two imperative sentences, two interrogative sentences, and two exclamatory sentences.

1. _____
2. _____
3. _____
4. _____
5. _____
6. _____
7. _____
8. _____

D. Write a short dialogue involving two people talking about pollution. Include at least one declarative, one imperative, one interrogative, and one exclamatory sentence.

LESSON 13
USE VARIETY: SENTENCE LENGTH

If you have too many sentences of the same length, your writing may be dull and repetitive. To keep your readers interested, **vary the lengths of your sentences**. Use a variety of short, medium, and long sentences. Long, descriptive sentences help readers picture what you are writing about and give them details about your topic. Short sentences can be used to grab readers' attention or show surprise or excitement. You can achieve sentence fluency by using a variety of sentence lengths in your writing.

A. **Rewrite the following paragraph, using a variety of short, medium, and long sentences to make it flow more naturally.**

I practised all week for today's track-and-field meet. I was nervous. I ran as fast as I could. I tried to catch up to the other runners. I was getting tired. I could feel myself sweating. I was breathing hard. I crossed the finish line. I came in third place.

_____
_____
_____
_____
_____

B. **Write a paragraph about a time when you worked very hard to achieve something, such as studying for a test or learning how to ice skate. Use a variety of sentence lengths in your paragraph.**

_____
_____
_____
_____
_____
_____
_____
_____

COMBINE SENTENCES: COMPOUND SENTENCES

A **compound sentence** is made up of at least two independent sentences. One way to make a compound sentence is to join the sentences with a conjunction, such as *and*, *or*, *but*, *so*, or *yet*. Notice that you need a comma before the conjunction.

For example: The tire on my bicycle is flat. I had to walk to school.
The tire on my bicycle is flat, *so* I had to walk to school.

You can use compound sentences to link related ideas or to vary the length of your sentences. This makes your writing more interesting for readers.

A. For each pair of simple sentences, circle the conjunction that could be used to make a compound sentence.

1. We arrived early for the appointment. We read some magazines in the waiting room.

    but   yet   so   or

2. I wanted to have spaghetti for dinner. My dad made hamburgers.

    or   and   but   so

3. We could go out for lunch. We could eat in the cafeteria.

    so   but   or   and

B. Join the sentences below with *and, or, but, so,* or *yet* to make a compound sentence.

1. The tornado tore through town. It damaged a lot of houses.

    _____

2. Fareed practised a lot for the tryouts. He did not make the soccer team.

    _____

C. Underline the compound sentences in the following paragraph:

I was really excited about the weekend. My mom and I were going to the art gallery, and I had been looking forward to it for weeks. My favourite artist was having an exhibit there. Her paintings are really beautiful, with lots of colours and interesting shapes. She mostly does landscapes, but sometimes she does portraits too. When we got to the art gallery, there was a huge lineup for tickets. We had bought our tickets in advance, so we could skip to the front of the line. The exhibit was amazing. After walking through the gallery, we went to the gift shop. My mom bought me a print of my favourite painting, and I put it up on my bedroom wall.

D. **Change the following pairs of sentences into compound sentences. Remember to add a comma and a conjunction.**

1. Imani invited seven people to her party. Only five people could come.

   _____
   _____

2. Aslan and Gita saw an owl perched on a tree by the side of the road. They stopped the car to get a closer look.

   _____
   _____

3. It might rain during the parade this weekend. It might be sunny.

   _____
   _____

4. Carla and I stayed up late working on our science project. We finished it in time for the science fair.

   _____
   _____

5. I woke up feeling sick this morning. I stayed home from school today.

   _____
   _____

E. **Write a paragraph about an event in Canadian history. Use at least three compound sentences in your paragraph. Remember that each part of a compound sentence can stand on its own as an independent sentence.**

_____
_____
_____
_____
_____
_____
_____
_____
_____

EXPAND SENTENCES: ADDING DETAILS

Writers expand simple sentences by **adding details** that give a clear picture of what they are writing about. Readers stay more interested if they can sense the who, what, where, when, why, or how of a sentence. One way to add details is to use adjectives and adverbs.

For example: The teacher asked me for the answer.
The *frustrated* teacher *sternly* asked me for the answer.
The *patient* teacher *politely* asked me for the answer.

In the examples above, each adjective and adverb is one word. Sometimes, adjectives and adverbs can be more than one word. These are called adjective phrases and adverb phrases. An adjective phrase is a group of words that functions as an adjective.

For example: The girl *by the window* was daydreaming.

An adverb phrase is a group of words that functions as an adverb.

For example: She hit the volleyball *with great force*.

Expanding sentences also helps you vary their length so that your sentences do not all sound the same.

A. **In the sentences below, underline the word that gives details about the subject (adjective). Circle the word that gives details about the verb (adverb).**

1. She quickly turned around and faced the angry dog.
2. The owl gracefully swooped down and snatched up the frightened mouse.
3. Felicia had carefully planned every detail of her party, and she was sure it would be a memorable event.

B. **Rewrite the sentences by changing the underlined details. On the lines provided, write each sentence with new details.**

1. <u>She</u> noticed a <u>yellow</u> kite in the <u>ancient oak</u> tree.

   _____

   _____

2. The <u>old</u> man with the <u>round</u> glasses was <u>walking slowly</u> toward <u>her</u>.

   _____

   _____

3. The <u>ambitious</u> politician argued her points with <u>passion</u> and <u>enthusiasm</u>.

   _____

   _____

C. Expand the sentences below by adding details. Rewrite each sentence, adding one or two adjectives or adjective phrases for the subject and one adverb or adverb phrase for the verb.

1. He walked home.

_____

2. She painted a picture.

_____

3. The fans cheered.

_____

4. Anita crossed the finish line.

_____

5. The woman looked at me.

_____

D. Rewrite the following paragraph. Expand the sentences by adding details.

Spring is my favourite time of year. I love watching people playing in the park. I love to see the flowers bloom. But, most of all, I love that it is warm outside and that the snow is gone. My sister and I like to play in the backyard. We take our dog for walks. Sometimes, we hang out with the kids next door.

_____

_____

_____

_____

_____

_____

_____

_____

_____

_____

_____

E. Look through some of your previous writing. Do any parts seem dull or lacking details? Try expanding the sentences to add details. Then, compare your new version with the previous one. How did adding details change the experience as a reader?

LESSON 16: EDIT SENTENCES: RUN-ON SENTENCES

A **run-on sentence** is two (or more) sentences that have been put together as one sentence without the correct punctuation or linking words.

For example: I worked hard on my essay my mom helped me a little. ✗

You can fix run-on sentences in two ways: 1) make two separate sentences; 2) use a comma and a linking word (*and*, *so*, *but*, *otherwise*, *although*) to make a compound sentence.

For example: I worked hard on my essay. My mom helped me a little.
I worked hard on my essay, although my mom helped me a little.

Avoid run-on sentences as you write by being aware of the parts of your sentences. If your sentence should have a pause but there is no punctuation to suggest one, you might have a run-on sentence.

A. **Identify the run-on sentences. Place an ✗ next to each run-on sentence and a check mark (✓) next to the sentences that are correct.**

   1. The newest member of the debating team kept interrupting people and refused to wait his turn. _____

   2. Zelda was not a fan of the cafeteria she started bringing her own lunch. _____

   3. Climate change is an important issue, but some people still don't take it seriously. _____

   4. Recycle your technology gadgets they can harm the environment. _____

B. **Correct these run-on sentences.**

   1. Karina wasn't wearing safety goggles the teacher asked her to put on a pair.

   _____
   _____

   2. The speakers talked about cyberbullying it is an ongoing problem.

   _____
   _____

   3. I want to become a doctor like my mom it takes a lot of hard work.

   _____
   _____

C. **In your own words, define a run-on sentence and identify the ways that you can fix one. Then, find an example of a run-on sentence from your own writing and correct it using each method you have identified.**

LESSON 17
KNOW COMPLETE SUBJECTS AND PREDICATES

A sentence has two parts: subject and predicate.

The **complete subject** includes all the words that tell who or what is doing something, or whom or what the sentence is about.

The **complete predicate** includes all the words that tell what the subject is doing or what it is like.

For example: The hot sun melted the ice-covered lake.
Complete subject: *the hot sun*
Complete predicate: *melted the ice-covered lake*

Being aware of the complete subject and predicate helps you understand how your sentence fits together when you write.

A. Underline the complete subject and circle the complete predicate in each sentence.

1. My older sister won first prize in the speech-writing competition.
2. Meghan and her friends thought of several ways to reduce their carbon footprint.
3. The neighbour's dog was digging in our yard again this weekend.
4. The hydroelectric dam disrupted the freshwater ecosystem around it.
5. Kyle's parents were proud of him for passing his geometry test.
6. The bold raccoon broke through the screen and started rummaging through the cupboards.
7. The ideal summer vacation starts with a jump in the lake.
8. All runners need to stretch properly before and after going for a run.
9. My dog and I play fetch in the park every day after school.
10. Canadians celebrate the birth of their nation every year on July 1.

B. Write a complete predicate for each complete subject.

1. The rowdy hockey fans _____.

2. A friendly stranger _____.

3. A crowd of people _____.

4. My new computer _____.

5. A cloud of smoke _____.

6. The angry goat _____.

C. Write several paragraphs about going on an adventure in the wilderness. Use adjectives and adverbs in your complete subjects and complete predicates. Afterwards, underline the complete subjects and circle the complete predicates.

LESSON 18
IDENTIFY WHO OR WHAT: SIMPLE SUBJECTS

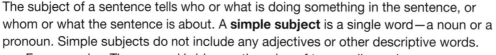

The subject of a sentence tells who or what is doing something in the sentence, or whom or what the sentence is about. A **simple subject** is a single word—a noun or a pronoun. Simple subjects do not include any adjectives or other descriptive words.
  For example: The covered bridge on the edge of town collapsed.

In the example, *bridge* is the simple subject. Notice that it does not include *The, covered,* or *on the edge of town*.

A. **Underline the simple subject in each sentence.**

   1. Innovative technology is changing the way we communicate.
   2. The injured hockey player on the opposing team had to sit out the rest of the game.
   3. The magnitude 7 earthquake on the west coast caused major damage in the surrounding areas.
   4. My favourite writer of all time came to speak at our local bookstore last week.
   5. The oldest book on the shelf was one that belonged to my great-grandmother.
   6. A good athlete is one who can accept defeat gracefully.
   7. The only problem with having a cat is when it scratches the furniture.
   8. Every bowl of cereal should be accompanied by a big glass of orange juice.

B. **Write a sentence about each topic below. Then, underline the simple subject in each sentence.**

   1. travelling to another country

   _____

   2. my favourite movie

   _____

   3. an after-school activity

   _____

   4. an event in my community

   _____

   5. the best meal I ever ate

   _____

LESSON 19
IDENTIFY THE ACTION: SIMPLE PREDICATES

A **simple predicate** contains only the verb from the action part of a sentence. Simple predicates do not include any adverbs or other descriptive words.
   For example: The colourful cardinal *flew* gracefully around the tree.

In the example, *flew* is the simple predicate. Notice that it does not include the descriptors *gracefully* or *around the tree*.

Simple predicates can sometimes be more than one word. Some verb forms are several words long.
   For example: She *was walking* to the store when she ran into her neighbour.
            Aaron *had met* her somewhere before.

A. Underline the simple predicate in each sentence.

   1. I went outside to enjoy the nice spring weather.
   2. The art exhibit is receiving rave reviews from critics and art lovers alike.
   3. After school, Zhang and Mariam started working on their science projects.
   4. The bake sale raised $200 for the homeless shelter.
   5. She had known Silas for seven years.
   6. The dogs were barking loudly at a strange noise outside the house.
   7. Carter was getting excited about the trip to Ottawa next week.
   8. The groundhog had been digging a huge hole in the middle of the vegetable garden.

B. Write three sentences about each topic below. Then, underline the simple predicate in each sentence.

   1. something that bothers me

   _____
   _____
   _____

   2. my ideal day

   _____
   _____
   _____

C. Without referring to the box at the top of this page, write a definition of simple predicates in your own words.

LESSON 20
IDENTIFY SENTENCE PARTS: DIRECT AND INDIRECT OBJECTS

An object is a noun or pronoun that receives the action in a sentence. There are two kinds of objects: **direct objects** and **indirect objects**.

A direct object receives the direct action of a verb.
 For example: Tyrese ate pancakes.

In this example, *ate* is the verb (the action). What did Tyrese eat? *Pancakes*. Therefore, *pancakes* is the direct object.

An indirect object indicates the person or thing to whom or for whom the action is being done.
 For example: Daniela gave me the book.

What did Daniela give? The book. Therefore, *the book* is the direct object. To whom did Daniela give the book? To me. Therefore, *me* is the indirect object.

A. Read the sentences below. Underline each direct object and circle each indirect object.

1. Cesar sent his grandmother a card.
2. My mom made me a costume.
3. The teacher showed us a documentary about the War of 1812.
4. A police officer wrote my dad a ticket for speeding.
5. Caitlin drew her sister a picture of a dinosaur.
6. The fierce dog showed me his teeth.
7. I offered my mom the spaghetti.

B. Write two sentences about two things that happened today. Include a direct object and an indirect object in each sentence. Then, underline each direct object and circle each indirect object.

 1. _____

 _____

 2. _____

 _____

C. Look at your favourite websites, books, or magazines, and find five sentences that contain an indirect object. Write down each sentence. Underline the direct object and circle the indirect object.

LESSON 21: RECOGNIZE INDEPENDENT AND SUBORDINATE CLAUSES

A clause is a group of words that has a subject and a predicate. There are two kinds of clauses. An **independent clause**, or main clause, expresses a complete thought and can stand on its own as a sentence.

For example: I dislike going to the dentist.

A **subordinate clause** has a subject and a predicate, but it cannot stand on its own as a sentence; it depends on an independent clause to make sense.

For example: because he always finds a cavity.

Subordinate clauses are important because they add information to a sentence. A subordinate clause begins with a conjunction, such as *after*, *before*, *because*, *unless*, *if*, *when*, *where*, or *whenever*. You can place a subordinate clause before or after the main clause. If you put it before the main clause, always follow it with a comma.

A. For each clause, identify whether it is a subordinate clause or an independent clause. Write *S* next to each subordinate clause and *I* next to each independent clause.

1. whenever I go outside _____
2. I borrowed some paper from Elijah _____
3. after she found her phone _____
4. this weather makes me tired _____
5. we learned about sustainable energy in class _____
6. unless it's not OK with your parents _____

B. Write an independent clause to go with each subordinate clause. Remember that independent clauses have a subject and a predicate and can stand alone as a sentence.

1. If I get home early, _____.
2. Whenever I think about climate change, _____.
3. Before I learned how to ride a bike, _____.

C. Write a sentence with one subordinate clause and one independent clause.

_____
_____
_____

D. Look through a magazine, book, or short story and find two examples of sentences that contain a subordinate clause and an independent clause. Write the sentences down. Underline the subordinate clause and circle the conjunction that introduces the subordinate clause in each sentence.

LESSON 22
COMBINE SENTENCES: COMPLEX SENTENCES

A **complex sentence** is a sentence that has an independent clause (or main clause), plus one or more subordinate clauses. You can join two separate sentences to form a complex sentence. One sentence remains an independent clause, while the other becomes a subordinate clause, giving more detail or description about the main sentence.

For example: I washed the dishes. My sister prepared dessert.
                I washed the dishes, while my sister prepared dessert.

You can join a subordinate clause to an independent clause with a conjunction, such as *although*, *after*, *before*, *because*, *since*, *unless*, *if*, *while*, *when*, or *whenever*. You can place the subordinate clause before or after the independent clause. If you place it before the independent clause, always follow it with a comma.

A. **Underline the independent clause in each complex sentence. Circle the word that links the independent clause with the subordinate clause.**

   1. The community needs to be consulted before the city can build a new park.
   2. Because they live close to the water, they sometimes experience flooding.
   3. We look at old photo albums whenever I visit my grandparents.
   4. We like to practise in the gym unless there is a class in there.
   5. Since there was heavy snow, people were advised to stay off the roads.
   6. Owen likes to go for a walk in the morning, if he can get up early enough.
   7. Although scientists were studying cholera in 1832, it wasn't until the following century that they discovered what caused it.

B. **Join each pair of independent clauses below to form a complex sentence using one of the linking words suggested. Choose which independent clause should become the subordinate clause, and add a linking word with the correct punctuation.**

unless   since

   1. I was running late for school. My mom gave me a ride.

   _____

   _____

   2. We practise hard on the debating team. We won't make it to the semifinals.

   _____

   _____

40   Build Sentences   Copyright © 2019 by Nelson Education Ltd.

C. **Complete each complex sentence by writing a subordinate clause to go with each independent clause. You can place the subordinate clause before or after the independent clause. Write your complete sentences on the lines provided. Remember to add a conjunction.**

1. We learned about the Upper Canada Rebellion.

   _____

   _____

2. Alex volunteers on Wednesdays.

   _____

   _____

3. Construction on the highway was finally finished.

   _____

   _____

D. **Write a paragraph or two about how you can make a difference in your community. Use at least two complex sentences in your paragraph.**

LESSON 23
RECOGNIZE CLAUSES: ADJECTIVE CLAUSES

> An **adjective clause** is a group of words that describes a noun or pronoun and has a subject and a verb. Adjective clauses can begin with relative pronouns, such as *who, whom, whose, that,* or *which*. They can also begin with relative adverbs, such as *when, where,* or *why*. Use an adjective clause to specify What kind? How many? Which one?
>
> For example: The clerk *who served us* was having a bad day.
>
> In the example, *who served us* is the adjective clause that describes the noun (*the clerk*). It describes which one, that is which clerk served us.

A. **Underline the adjective clause in each sentence.**

1. The time when I saw a caribou is one of my favourite memories.
2. Spending too much time online is the reason why I was grounded.
3. The woman whose purse was stolen is filing a police report.
4. The small town where I grew up is near Parry Sound.
5. I have a guitar that is more than sixty years old.

B. **Use your imagination to fill in the missing adjective clauses in the sentences below.**

1. Students _____ do well in school.
2. The fruit _____ had gone bad.
3. The house _____ is for sale.

C. **Write a short paragraph about where you grew up. Include at least two sentences that have adjective clauses. Underline each adjective clause.**

_____
_____
_____
_____
_____
_____

D. **Look through a book you are currently reading. Find five sentences with adjective clauses and write them down. Underline each adjective clause and explain what question the clause answers.**

RECOGNIZE CLAUSES: ADVERB CLAUSES

An **adverb clause** is a group of words that describes a verb, adjective, or another adverb. Adverb clauses are different from adverb phrases because they contain a subject and a verb.

Adverb clauses usually answer the questions *Where?* (place), *When?* (time), *Why?* (reason), or *How?* (under what condition). They always begin with a conjunction, such as *after*, *although*, *as*, *because*, *before*, *if*, *once*, *since*, *so*, *soon*, *unless*, *until*, *when*, *whenever*, and *while*.

For example: She planted the beans *where they would get a lot of sun.* (place)
I slipped and fell *when I was walking to school.* (time)
We came home early *because it started to rain.* (reason)
You will be late *unless you leave right now.* (condition)

Adverb clauses are kinds of subordinate clauses. They can go at the beginning or end of the sentence. If they go at the beginning, they must be followed by a comma.

A. **Underline the adverb clause in each sentence. Circle the conjunction that connects it to the main clause.**

1. I had to borrow my friend's phone because I forgot mine at school.
2. Luca will spend every waking hour practising trombone until he plays at the concert on Friday.
3. After I watched the documentary, I became passionate about fighting climate change.
4. Carmen browsed the library's history section while she waited for her mother to pick her up.
5. Since it was getting late, the Guptas decided to stop at a hotel for the night.
6. I will go to the party this weekend unless I have too much homework.

B. **For each topic, write a sentence that contains at least one adverb clause.**

1. Topic: new technology

_____

_____

2. Topic: health

_____

_____

C. **Look through some books in your home and find five adverb clauses. Write them down and underline each adverb clause. Then, write how you knew it was an adverb clause.**

LESSON 25: EDIT SENTENCES: SENTENCE FRAGMENTS

A **sentence fragment** is a group of words that is punctuated like a sentence, but is missing a subject or a verb.

    For example: Sneezing loudly in the middle of the play. ✗ (missing subject)
                      Boxes of food for the homeless shelter. ✗ (missing verb)

To fix this type of sentence fragment, add the missing subject or verb.
Another cause of sentence fragments is the incorrect use of a subordinate conjunction.
    For example: Although she told him she would be late. ✗

In this example, the subordinate conjunction *although* needs to connect to an independent clause. It cannot stand on its own because it is not a complete thought. To fix this type of sentence fragment, remove the subordinate conjunction, or connect it to an independent clause.
    For example: She told him she would be late. ✓
                      Although she told him she would be late, he arrived early. ✓

A. Place an *X* next to each sentence fragment below.

1. Until the tornado passed and people began to assess the damage.
2. Because every day since she joined the track team she became stronger and more confident.
3. It took a long time for Yves to convince his teacher to let him retake the exam.
4. A flock of geese flying in a V-shape overhead.
5. Galen finally found out what was making the scratching noise in his walls at night.
6. Other forms of sustainable energy, such as solar power, wind power, and hydroelectricity.

B. Correct the sentence fragments. Write the complete sentences on the lines provided.

1. Determined to find the person who stole her laptop from the library.

    _____
    _____

2. The best way to get better at playing the violin.

    _____
    _____

C. Sentence fragments are used frequently in advertisements. Look online or in magazines to find examples of sentence fragments in advertising. Correct the sentence fragments that you find. Then, in writing, explain how the images or other information in the ad helped you to make your correction. How did the tone of the ads change when you made complete sentences?

44   Build Sentences   Copyright © 2019 by Nelson Education Ltd.

LESSON 26: EDIT SENTENCES: COMMA SPLICES

A **comma splice** occurs when a comma is used to connect two independent clauses. This is an incorrect use of a comma.

    For example: She went to Europe to study art, it was an unforgettable experience. ✗

There are three different ways to fix a comma splice:

1. Make two separate sentences.
       For example: She went to Europe to study art. It was an unforgettable experience. ✓
2. Add a coordinating conjunction.
       For example: She went to Europe to study art, *and* it was an unforgettable experience. ✓
3. Replace the comma with a semicolon.
       For example: She went to Europe to study art; it was an unforgettable experience. ✓

A. Underline the sentences with comma-splice errors in the following paragraph. Then, rewrite those sentences to correct the comma-splice errors. Write the new sentences on the lines provided.

Esther Brandeau was a brave woman, she was the first Jewish person to set foot on Canadian soil. Esther sailed to Québec in 1738, she said her name was Jacques La Fargue. At that time, Jewish people were not allowed in Canada. Her identity was discovered, she refused to convert to Catholicism.

_____

_____

_____

_____

_____

B. Fix the comma-splice error below in two different ways. Write the corrected sentences on the lines provided.

It was a major earthquake, many families had to be relocated.

_____

_____

C. Compare the effects of the two different sentences you wrote for Exercise B. Write a short statement about which is more effective and why.

SECTION REVIEW

A. **For each sentence, write on the line provided whether it is a statement, command, question, or exclamation.**

   1. Lady Elizabeth Simcoe painted many watercolours. _____

   2. Write a thank-you note to your uncle. _____

   3. Trust me, I am never making that mistake again! _____

   4. Did anyone understand the geometry homework last night? _____

B. **Rewrite the following passage using a variety of sentence lengths:**

   We walked to the seniors' centre. It was my first volunteer shift. I went into Mrs. Yang's room. I liked talking to her. She told me about her past. She was a violinist, she played in an orchestra, she had five children, she had pictures of them.

   _____

   _____

   _____

   _____

C. **Correct these run-on sentences by either adding a comma and a conjunction or by separating them into two sentences.**

   1. I wanted to learn about journalism my aunt let me job-shadow her.
   _____

   2. We couldn't believe the amount of litter on our street it made us want to take action.
   _____

   3. The library was loud we had to find somewhere else to study.
   _____

   4. The yard sale was a huge success we raised a lot of money.
   _____

D. **Underline the complete subject and circle the complete predicate in each sentence.**

   1. A large group of tourists gathered around to take pictures of the war memorial.

   2. My dad took me out on the sailboat for the first time this summer.

   3. The bus full of travellers stopped suddenly to avoid hitting the turtle in the middle of the road.

   4. My dog never seems to realize that his tail is actually attached to the rest of his body.

E. **Underline the simple subject and circle the simple predicate in each sentence.**

1. The woman who lives next door complains if she hears a pin drop.
2. My music teacher told me to practise playing the piano every night.
3. My class started a petition for the city to ban water bottles.

F. **Underline the direct object and circle the indirect object in each sentence.**

1. My teacher gave me some advice about writing a good essay.
2. Tunde sent her grandmother a card for Mother's Day.
3. My father wrote me a note when I was late for school.

G. **Identify whether each clause is an independent clause or a subordinate clause. Write *I* next to each independent clause and *S* next to each subordinate clause.**

1. until I saw my mother standing at the window _____
2. we walked across the baseball field _____
3. the pipes in the kitchen burst _____
4. whenever she gets home early _____

H. **Underline the independent clause in each complex sentence.**

1. Whenever we heard that song on the radio, we started singing at the top of our lungs.
2. Laura took notes while I read my speech aloud.
3. If we win the next competition, we'll make it to the final round.
4. I closed my eyes until the scary part was over.

I. **Underline the adjective clause in each sentence.**

1. The man who found my phone refused to accept a reward.
2. We are going to visit the house where my grandmother grew up.
3. The time when I sprained my ankle was my first visit to the hospital.
4. The vegetables that I planted are starting to sprout.

J. **Fix these sentence fragments and comma-splice errors.**

1. As soon as she left the movie theatre.

   _____

2. Crowds of angry hockey fans.

   _____

3. Realizing I had overslept.

   _____

4. Adley wrote a song about his school, he sang it at the end-of-year talent show.

   _____

K. **Reread the sentences in Exercise J and choose one to use as the basis of a story. Include at least three compound sentences.**

L. Write one or more paragraphs about the effects of new technology. Focus on two devices you use or have seen. Include at least three complex sentences.

Know Capitalization and Punctuation

There are many rules in every language, from how to spell a word to how to end a sentence. Most of the rules are unbreakable —if you break them, you're just plain wrong. Your message will be difficult to understand, and people might not take your writing seriously.

Some of the rules, such as when to use a comma, have a little more flexibility. But, like learning a new game, you need to know the rules before you can begin playing.

In this section, you will get to know these rules so that you can guide readers through your writing just the way you want to.

> "I'm tired of wasting letters when punctuation will do, period."
> — Steve Martin

LESSON 27: USE CAPITALS: A VARIETY OF CAPITALIZATION

We use **capitals** for many kinds of words, including the following:
- the first word of a sentence
- the first word and other important words in a heading, subheading, or title (song, poem, book, or story)
- days of the week, months of the year, holidays, and special days
- important and specific places, events, historical periods, and documents
- the planets and other specific objects in the universe
- proper nouns (such as the names of people, organizations, and schools)
- abbreviations (such as days and months, forms of address, places, and geographical features)
- important words in the names of awards, such as Volunteer of the Year

We also use capital letters for adjectives formed from proper nouns, such as *Chinese* and *Canadian*.

A. Read each set of words. Underline the set that needs capital letters.

1. cape breton island   the farthest island
2. the niagara treaty of 1764   a long-lasting peace treaty
3. the commanding general   gen. james wolfe
4. the north star   the brightest star
5. six national holidays   national acadian day

B. In the sentences below, underline each letter that should be capitalized.

1. the québec act of 1774 guaranteed religious freedom for the roman catholics in that province.
2. the city of brantford is named after joseph brant, who was a political leader of the mohawk nation.

C. Rewrite each sentence, using capitals where necessary.

1. africville was an african-canadian community near halifax.

_____

_____

2. the africville genealogy society published a book called the spirit of africville.

_____

_____

3. another book, called last days in africville, was nominated for the book of the year for children award.

_____

D. Write two narrative paragraphs about a historical event or an event that is important to you. Make sure to mention the people involved, the date(s), and the place(s). Use capitals where necessary.

LESSON 28

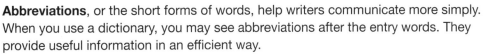
USE A DICTIONARY: ABBREVIATIONS

Abbreviations, or the short forms of words, help writers communicate more simply. When you use a dictionary, you may see abbreviations after the entry words. They provide useful information in an efficient way.

For example: **sustain** *v.* keep up, or continue for a period of time
sustainability *n.* the ability of something to be continued
sustainable *adj.* that which can be continued
sustainably *adv.* in a way that allows something to be continued

The abbreviations *v.*, *n.*, *adj.*, and *adv.* indicate parts of speech: *verb, noun, adjective,* and *adverb*.

You might also see these abbreviations in a dictionary:

art. meaning "article" *contr.* meaning "contraction" *colloq.* meaning "colloquial"
prep. meaning "preposition" *sing.* meaning "singular" *pl.* meaning "plural"
conj. meaning "conjunction" *abbrev.* meaning "abbreviation"

A. Choose the answer to each question below from the list underneath it. Underline your choice.

1. Which of the following does *not* indicate a part of speech?

   a) *abbrev.*
   b) *art.*
   c) *adj.*
   d) *adv.*

2. Which abbreviations would you see after the word bacteria?

   a) *n., v.*
   b) *contr., pl.*
   c) *abbrev., sing.*
   d) *n., pl.*

3. Which word is a describing word that relates to plants or animals?

   a) organism *n.*
   b) organic *adj.*
   c) organize *v.*
   d) org *abbrev.*

B. Complete the sentence, using one of the following words on each of the lines provided. Use the dictionary abbreviations to help you choose the correct word.

from *prep.*   **decompose** *v.*   **also** *adv.*   **fungi** *n., pl.*   **an** *art.*   **because** *conj.*

The death of (article) _____ animal means life for (plural noun) _____,
(conjunction) _____ as fungi (verb) _____ the organic material
(preposition) _____ the animal, they (adverb) _____ consume
nutrients from it.

C. Look through a dictionary and choose a word that can be used as a noun, a verb, and an adjective (for example, *discount*). Write three sentences with your chosen word in each of its forms.

IDENTIFY SHORT FORMS: ABBREVIATIONS

Abbreviations, or the short forms of words, help writers to communicate more quickly. Abbreviations can also make writing easier to read because they replace long and/or difficult words.

It is important to know when abbreviations are appropriate, however. Consider the audience for your writing and the tone of your piece. If your audience might not know the long form of an abbreviation, you should not use the short form. If you are writing something that requires a formal tone, you should avoid abbreviations, just as you should avoid contractions.

For example: The following email to a teacher would not be appropriate:
Hi Mr. Ye. Can u send me todays homework pls? sry I forgot to bring it home from school lol btw I forgot to tell u I want to sign up for volleyball. ty! Selma

A. Rewrite each sentence, using more formal language to replace each abbreviation.

1. I'm 100% lol irl rn

   _____

2. Mte! That's why ur my bff

   _____

3. tbh idc what we do tonight

   _____

B. Write two sentences. The first sentence should include several abbreviations. The second sentence should be a more formal version of the first sentence. At the end of each sentence, identify your audience.

_____

_____

_____

_____

C. Rewrite the informal email in the information box at the beginning of this lesson using more formal language. Don't just replace the abbreviations, though. Look carefully at each sentence to decide if a more formal word or phrase can replace an informal one.

LESSON 30
USE VARIETY: COMMAS

Writers use **commas** in their sentences in the following situations:

- after *Yes*, *No*, *Well*, or a name at the beginning of a sentence
- after each item, except the last one, in lists of three items or more
- after who said dialogue, if the dialogue comes second in a sentence
- before the last quotation mark, if the dialogue comes before who said it
- between two parts of a compound sentence

    Remember that a compound sentence is two sentences joined together with a word such as *or*, *and*, or *but*.

- after a subordinate clause, if it comes before an independent clause in a sentence

    Remember that a *subordinate clause* has a subject and a predicate, but it must be combined with an independent clause. Subordinate clauses start with words such as *if*, *although*, *once*, *since*, *after*, *before*, *because*, *unless*, *until*, *when*, *where*, and *while*.

A. Rewrite each sentence, placing commas in the appropriate places.

1. "Let's review conduction convection and radiation" Mr. Pinder said.

    _____

    _____

2. Camilla asked "Mr. Pinder when we talk about conduction can you please give us a definition?"

    _____

    _____

3. "Yes I will" he answered "as soon as Chantal Samson and Inez stop talking!"

    _____

    _____

B. Add commas to the following paragraph where they are needed:

"Well I'd like to start today's meeting by thanking everyone for coming" the volunteer coordinator said. She continued "I know it's a beautiful day and we would all love to be outside rather than here. Because the weather is so nice I'll try to get through our agenda as quickly as possible. Is it OK if we skip our break or will people need one in an hour? When we finish discussing this first item we can decide."

C. Write a conversation among three people discussing their plans for the weekend. Each person should speak a minimum of two times. Include one compound sentence and one subordinate clause that comes before an independent clause. Use commas where necessary.

LESSON 31: PUNCTUATE DIALOGUE: QUOTATION MARKS

When you write dialogue, use **double quotation marks** (" ") around what someone says.
   For example: Pradeep told the audience, **"I'm going to read my new story."**

Use **single quotation marks** (' ') around the title of a story, poem, or song *within* someone's speech.
   For example: Pradeep continued, **"The story is called 'Lost in the City.'"**

Also use single quotation marks inside double quotation marks if the speaker quotes someone.
   For example: He said, **"My grandfather once told me that 'you can feel lost even in your own hometown.'"**

Notice that the period appears before the last single quotation mark, which appears before the last double quotation mark. The same rule applies to commas.

A. **Rewrite each sentence, adding a title as indicated. Add the correct quotation marks.**

   1. Adi affirmed, "[poem title] is my all-time favourite poem."

   2. "I think you would really like the short story called [short story title]," Eden said.

B. **In the dialogue below, write single and double quotation marks wherever they are missing.**

   I knew I would never feel lost again was my favourite line in your story, Pradeep, Myra said.

   Pradeep replied, Thanks! I originally wrote I knew I would never *be* lost again, but I changed it.

   Myra commented, Your story reminded me of the song Lost Stars.

C. **Rewrite each sentence, adding single and double quotation marks where needed.**

   1. Falling Down is my least favourite song right now, said Arianna.

   2. If I hear the song Go Around one more time, I think I'll scream, she added.

   3. Marie-Louise agreed, Yes, I'd much rather listen to Frantic.

D. Write a dialogue between two people discussing what to call the TV show they have created. Use single and double quotation marks correctly.

LESSON 32
SHOW POSSESSION: APOSTROPHES

Use an **apostrophe** (') when you want to show possession or ownership in your writing. Follow these rules:

For singular nouns (including those ending in *s*), add apostrophe + *s*.
   For example: the girl's bicycle / the walrus's skin

For plural nouns ending in *s*, add only an apostrophe.
   For example: the trains' engineers / the Greens' vacation

For plural nouns that do not end in *s*, add apostrophe + *s*.
   For example: the children's playroom

For two or more nouns that own something *together*, make only the *last* noun possessive.
   For example: Jen and *Jill's* puppies were checked by the vet.
                  Jen and Jill own all the puppies.

For two or more nouns that *each* owns something *separately*, make *each* of the nouns possessive.
   For example: *Jen's* and *Jill's* puppies were checked by the vet.
                  Jen owns some of the puppies and Jill owns some of the puppies.

A. **Underline the correct ending for each item. Refer to the rules above to help you.**

1. Marco and Kevin each own safety glasses. They are—
   a) Marco's and Kevin's glasses.
   b) Marco and Kevin's glasses.

2. Bonnie and Nuala share tools. They are—
   a) Bonnie's and Nuala's tools.
   b) Bonnie and Nuala's tools.

3. The grinder and the drill have separate instructions. They are—
   a) the grinder and the drill's instructions.
   b) the grinder's and the drill's instructions.

B. **In the sentences below, find the nouns that are missing an apostrophe to show possession. Write the noun, with the apostrophe added, on the line beside the sentence. The number in parentheses tells you how many possessives there are in each sentence.**

1. Last weekend I sailed on my aunts boat for the first time. (1)_____

2. The boats hull is blue and the decks surface is white. (2) _____ _____

3. The masts sails flapped in the wind. (1) _____

4. We heard a loons call and an albatross cry as we sailed (2)_____ _____

C. **Think of three plural nouns for items in your home that are shared and three more plural nouns for items that are used by separate people. Write a sentence for each item that includes a subject with at least two nouns.**

LESSON 33: JOIN INDEPENDENT CLAUSES: SEMICOLONS

A clause is a group of words with both a subject and a predicate. An independent clause can stand on its own as a sentence. It expresses a complete thought.

For example: Friends of the Earth believes we deserve a clean environment. The organization thinks the government should help to make this happen.

In the example sentences, *Friends of the Earth* and *The organization* are the complete subjects. The predicates are *believes we deserve a clean environment* and *thinks the government should help to make this happen*.

Sometimes, writers join two independent clauses to form one sentence. They use a **semicolon** (;) between the clauses to show that the two thoughts are closely related, but they are also complete on their own.

For example: Friends of the Earth believes we deserve a clean environment; the organization thinks the government should help to make this happen.

Notice that the word after the semicolon does not begin with a capital letter.

A. In each sentence, underline the complete subjects and circle the complete predicates. Remember that there is one of each on both sides of the semicolon.

1. Many environmentalists oppose oil pipelines; they believe pipelines damage water sources and the land.

2. Some Indigenous communities do not want oil pipelines on their land; pipelines can disturb sacred places and contaminate fish.

B. Join the two independent clauses below to form one sentence. Write the sentence on the lines provided. Use a semicolon between the two clauses.

1. Humans use water for many purposes besides drinking and cleaning. We also use it to produce electricity and to water crops.

_____
_____

2. Canada's mines produce gold, silver, and diamonds. They are also known for nickel, iron, and coal.

_____
_____

C. Find a website or book that provides information on one of the topics in Exercise A or B. Look for two independent clauses that contain related thoughts. Rewrite the two clauses as one sentence, joined with a semicolon.

LESSON 34: SEPARATE TITLES AND SUBTITLES: COLONS

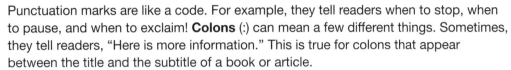

Punctuation marks are like a code. For example, they tell readers when to stop, when to pause, and when to exclaim! **Colons** (:) can mean a few different things. Sometimes, they tell readers, "Here is more information." This is true for colons that appear between the title and the subtitle of a book or article.

For example: "The Best Road Trip Ever: Three Weeks in Newfoundland"

In the example, the title is "The Best Road Trip Ever." The colon means, "Here is more information about the best road trip ever." The subtitle is "Three Weeks in Newfoundland," which tells where the road trip was and how long it lasted. This is the "more information" that the colon promised.

Notice that the first word after the colon (the first word in the subtitle) is capitalized. When a title includes quotation marks, a question mark, or an exclamation mark, place the punctuation before the colon.

For example: "Bravo!": A Guide to Performing Outstanding Monologues

A. Match each title on the left with a subtitle on the right by drawing a line between them. Make sure the subtitle gives more information about the title. Write the pairs on the lines below, using a colon to separate the title and subtitle.

"Good Try!"   Local Support Resources

Take Charge of Your Health   How to Be a Supportive Teammate

Help Is Just a Phone Call Away   A Nutrition Guide for Teenagers

1. _____
2. _____
3. _____

B. For each title, write a subtitle that provides more information. Don't forget the colon.

1. Why Am I Stressed? _____
2. That's Not OK _____
3. "Goooaaal!" _____
4. Smoothie Smarts _____

C. Create a title for each of the following items. Add a subtitle.

1. a web article about your favourite actor _____
_____

2. a report for science class about fungus _____
_____

ADD LESS IMPORTANT INFORMATION: PARENTHESES

When you write a text, not all of the sentences have the same level of importance. Some sentences might not be absolutely necessary for understanding your main idea. They might just provide an interesting fact or help to make a detail clearer. If you can remove the sentence without altering the meaning or point of your writing, then consider placing it in **parentheses ( )**.

You can use parentheses to show readers which sentences are less important than others. Parentheses come in pairs: one parenthesis comes before the less important sentence and one comes immediately after it.

For example: Making these cupcakes will take 30 minutes. It will take longer if this is your first time.

Making these cupcakes will take 30 minutes. (It will take longer if this is your first time.)

Notice that the end punctuation mark of the less important sentence comes *inside* the second parenthesis. Do not overuse parentheses in your writing! Used sparingly, they are a stylistic device that can give voice to your writing.

A. **For the group of sentences below, decide if the parentheses around the second sentence are used correctly. Underline *Yes* or *No*.**

Many people who came to Canada in the 1700s had never experienced harsh winters. (They were not prepared for the cold and the snow.) As a result, many people died.   Yes   No

B. **In the paragraph below, write parentheses around the sentences that are less important. Remember, if you can remove the information without altering the meaning of what remains, you should be able to place parentheses around the sentence.**

First, preheat your oven to 350 °C. Remember to make sure it's empty first! Then, gather all of the ingredients. Feel free to replace the walnuts with pecans or almonds. Grease a muffin pan with butter and set it aside. In a large bowl, beat the butter, and then add the eggs and sugar. Beat until smooth. In a medium bowl, combine the flour, baking powder, and salt. By now, your oven should be at the correct temperature. Add the dry ingredients and the milk into the egg mixture. Stir until the dry ingredients are just incorporated. If you stir the batter too much, your muffins will not be light and fluffy. Add the nuts into the batter. Spoon the batter into the muffin pan, and then bake for 25 minutes.

C. **Choose an activity, such as making your favourite meal or packing for a camping trip. Write instructions for someone who has never completed this activity before. Place parentheses around information that is less important than the necessary instructions but is still interesting or useful to include.**

LESSON 36: GUIDE READERS: A VARIETY OF PUNCTUATION

When you write, you need to choose your **punctuation** as carefully as your words. Punctuation marks guide readers through your sentences and help them understand what you mean. Make sure you understand the following ways to use punctuation:
- double quotation marks around dialogue
- single quotation marks within dialogue around quotes and the titles of stories, songs, and poems
- apostrophes to show possession or ownership
- semicolons to join two independent clauses with closely related ideas
- colons to separate titles from subtitles
- parentheses around sentences that contain less important information

A. **Each item below has two parts. Decide if you need a semicolon or a colon to join the parts.**

1. Paradise in Alberta___ Banff's Year-Round Appeal
2. Prem went skiing while Asa tried snowshoeing___ their parents enjoyed the hot springs.
3. In the summer, you can explore the park on foot or on a bike___ you can even explore it by helicopter!
4. Our Trip Blog ___ Best Vacation Ever!

B. **Rewrite the following sentence, using the correct punctuation.**

Big, Big Snow is Essie and Yusuf's favourite song from the festival. said their mother.

_____

_____

C. **Add the missing punctuation to the dialogue below:**

We saw a huge drum and dancers in regalia at the festival said Essie excitedly.

We also listened to a storyteller added Yusuf She told a story called The Last Nakoda.

The siblings had thoroughly enjoyed the festival dancers they had loved the music, too. Yusuf's favourite new food was breakfast bannock. The brother and sisters vacation was off to a great start the rest of their planned activities also sounded wonderful.

Tomorrow we're going hiking, right asked Essie

Yusuf answered That's right. Tomorrow is the hike. Wednesday is fishing day!

D. **Write a short story about a family visiting a new place. Include some dialogue, some titles, and a few less important sentences. Use the punctuation you have practised.**

SECTION REVIEW

A. Rewrite each sentence, using capital letters where necessary.

1. on april 17, 1982, the constitution act came into effect.

   _____

2. Queen elizabeth II signed the document in the city of ottawa.

   _____

B. When should you use social media abbreviations? Read each situation listed below. Write "Yes" on the line if social media abbreviations are appropriate in the situation, and write "No" on the line if they are not appropriate.

1. My audience might not know the long forms of the abbreviations I want to use. _____

2. My purpose is to apply for a volunteer position at a summer day camp. _____

3. My audience is my best friend, and my purpose is to joke with him. _____

C. Rewrite this paragraph, adding the missing commas.

"Well being a good actor involves a good memory the willingness to take risks and a love of good stories" Aya told the reporter. "You can attend theatre school but sometimes experience is more valuable than knowledge."

_____

_____

_____

_____

D. For each sentence, add single and double quotation marks where they are missing.

1. I think we should call our skit A Day in the Life of a Mountain Climber, Prem said.

2. Tackling Mount Everest has a nice ring to it, too, suggested Yazmeen.

3. How about calling it Climbing to Success or something? Oskar added.

4. Looks like we'll need a vote! Prem said.

E. Complete each sentence by writing the correct possessive on the line provided. Think about whether ownership is shared or separate in each situation.

1. This trail is for both walkers and bikers. It is a _____ trail.

2. This water fountain is for dogs. That one is for people. They are the _____ fountains.

3. My brother Andreas and my sister Gert each has a scooter. They are _____ scooters.

F. **Add a semicolon to correct each run-on sentence.**

 1. Olivia has taken several good photographs lately she especially likes the close-ups of flowers.

 2. She is thinking about entering her photos in a contest the winner receives a new camera!

 3. Olivia asks her aunt which photos she should enter she respects her aunt's opinion on photography.

G. **Pair each title on the left with a subtitle on the right by drawing a line between them. Make sure the subtitle gives more information about the title. Then, write the pairs on the lines below, using a colon to separate the title and subtitle.**

Thunderstorms	Three Ways to Produce Heat
Gassing Up	Nature's Convection Ovens
Hot, Hot, Hot!	Our Contribution to the Greenhouse Effect

 1. _____

 2. _____

 3. _____

H. **In the following paragraph, write parentheses around the sentences that are less important:**

For our experiment, we wanted to find the best way to keep certain foods fresh. Canadians waste a lot of food every year because it goes bad. We chose to use strawberries and various kinds of containers. We used four kinds of containers, but you can use more or fewer. We put the same number of strawberries in each kind of container and put all the containers in the fridge. We checked each container every day to see how fresh the strawberries were. You can use another type of food, as long as it's one that doesn't stay fresh for very long. We recorded the results in a log, and then turned the results into a graph. Our conclusion is that a lettuce keeper is the best container for keeping strawberries fresh. Another conclusion we came to is that handling rotten strawberries is disgusting!

I. **In each sentence, underline the punctuation marks that are used incorrectly. Then, rewrite the sentence, using the correct punctuation.**

 1. 'I've decided that I want to be a poet.' announced Oban.

 _____

 2. He added, 'I read an article called "How to Become a Poet, 10 Simple Steps": it sounds pretty easy.'

 _____

 3. I asked "Have you read the poem (I Lost My Talk) by Rita Joe?"

 _____

 4. Oban answered. "No: I haven't. I'll look for it in the library tomorrow."

 _____

J. Write a short dialogue between a reporter and a musician talking about a popular community music festival. Use capitals, commas, and quotation marks correctly.

K. Write two paragraphs giving advice to a friend who wants to learn how to do something. Include at least one sentence in which a semicolon separates two independent but related clauses.

GRASP GRAMMAR AND USAGE

Grammar is the guts of how language works. You might not know all the rules by name, but you use them whenever you write or speak.

Without grammar, writing would just be a mixed-up jumble of words. Grammar tells us what should come first in a sentence, what should come next, and how these two things are related. Without the right grammar, your writing will sound wrong, even if you're not sure why.

In this section, you will learn how paying attention to grammar can improve your writing.

> "... Prose gets its shape and strength from the bones of grammar."
> — Constance Hale

LESSON 37: NAME THE PERSON, PLACE, THING, OR IDEA: NOUNS

A **noun** is a word that names a person, place, thing, idea, or feeling.

A common noun names a general person, place, thing, idea, or feeling.
   For example: author (person); gymnasium (place); battery (thing); culture (idea); love (feeling)

A proper noun names a specific person, place, or thing. Proper nouns always have capital letters.
   For example: Mrs. Aranha (person); Manitoba (place); Canadian Space Agency (thing)

A collective noun is a word for a group of people or things.
   For example: a *family*; the students' *council*; a lacrosse *team*

Knowing about nouns helps you to choose the best ones for your purpose and to use them correctly in sentences.

A. Underline the nouns that name a thing, an idea, or a feeling.

assistant	cruise	detective	reflection	laptop
lowlands	woman	prairie	referee	field
employee	sediment	arcade	Europe	guilt
citizen	lifestyle	forest	mechanic	curiosity

B. Identify each noun as a common noun, proper noun, or collective noun.

1. team _____
2. restaurant _____
3. Moncton _____
4. document _____
5. community _____
6. bravery _____
7. Jupiter _____
8. audience _____

C. For each type of noun, write an example on the line provided.

Common Noun
person: _____
place: _____
thing: _____
idea: _____

Proper Noun
person: _____
place: _____
thing: _____

Collective Noun
people: _____
animals: _____

D. For each noun listed on the left, identify the type.

	Common, Collective, or Proper?	Person, Place, Thing, or Idea?
1. trust	_____	_____
2. Kluane Lake	_____	_____
3. family	_____	_____
4. ticket	_____	_____
5. Dr. Singh	_____	_____

E. Write one or more paragraphs on a topic related to a historical site you have visited or would like to visit in Canada. Use a variety of nouns in your writing, including proper nouns and collective nouns.

_____
_____
_____
_____
_____
_____
_____
_____
_____
_____
_____
_____
_____
_____
_____
_____
_____
_____

F. Make a chart for sorting nouns that you find in song titles. Your row headings might be *Common* and *Proper*, and your column headings might be *People*, *Places*, *Things*, and *Ideas*. Write a summary statement about the kinds of nouns that come up the most regularly and why.

SHOW OWNERSHIP: SINGULAR POSSESSIVE NOUNS

Possessive nouns are a concise way of showing ownership. For example, instead of writing "the mayor of the city," you could write "the city's mayor."

A **singular possessive noun** shows that one person, place, thing, or idea owns or has something. To make a singular noun possessive, you simply add an apostrophe (') followed by the letter *s*.

For example: the *mayor's* office; *Dawson City's* history; the *fox's* den; *time's* passing

If a singular noun already ends in *s*, you still just add an apostrophe (') plus *s* to make the possessive.

For example: the *boss's* problem; *Thomas's* phone; the *mattress's* filling

A. Rewrite each phrase below, using a singular possessive noun.

1. theme of the party _____
2. seats of the bus _____
3. resources of Earth _____
4. pen belonging to Marcus _____
5. climate of Alberta _____
6. laptops belonging to the library _____

B. For each sentence, underline the correct word in parentheses.

1. The (leaf's / leafs) edges are speckled with brown spots.
2. My grandmother grows (plants / plants') in her greenhouse.
3. That (valley's / valleys) homes are in danger of mudslides.
4. My (inboxes / inbox's) files are not appearing.
5. We use a variety of (instruments / instruments') in science class.
6. What is more fascinating than a (butterfly's / butterflies) wings?
7. Leafy (vegetable's / vegetables) are a good source of minerals and vitamins.

C. Write a sentence using the singular possessive form of the noun *Canada*.

_____

D. Choose four nouns: a person, place, thing, and idea. For each noun, write a sentence using the singular possessive form.

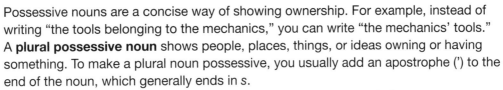

LESSON 39: SHOW GROUP OWNERSHIP: PLURAL POSSESSIVE NOUNS

Possessive nouns are a concise way of showing ownership. For example, instead of writing "the tools belonging to the mechanics," you can write "the mechanics' tools." A **plural possessive noun** shows people, places, things, or ideas owning or having something. To make a plural noun possessive, you usually add an apostrophe (') to the end of the noun, which generally ends in s.

For example: *teachers'* parking spaces (the parking spaces belong to the teachers)
shelves' contents (the contents belong to the shelves)
teams' schedules (the schedules belong to the teams)
cherries' flavour (the flavour belongs to the cherries)

Possessive nouns always have an apostrophe. If you add an apostrophe to a noun, make sure you are doing so to show possession. Plural nouns that are not possessive do not need an apostrophe.

A. Write the possessive form of these plural nouns.

1. groups _____
2. Manitobans _____
3. knives _____
4. turkeys _____
5. sandwiches _____
6. provinces _____

B. Decide if the underlined possessive noun is singular or plural. Write *S* for singular or *P* for plural on each line below.

1. The <u>scientists'</u> discovery was fascinating. _____
2. We got lost in the <u>forest's</u> many trails. _____
3. The <u>bears'</u> dens are sheltered and well hidden. _____
4. The <u>clubs'</u> addresses are on their web pages. _____
5. The <u>document's</u> title is too long. _____
6. The <u>earthworms'</u> castings add nutrients to the soil. _____

C. Replace the phrase in parentheses with its possessive form.

1. The _____ frightened the small children. (barks of the dogs)
2. The abandoned barn was in the _____. (paths of the tornados)

D. Look through some magazines and books for examples of plural possessive nouns. Write an explanation of how to recognize plural possessive nouns and when to use them in your own writing.

LESSON 40: USE IRREGULAR PLURAL POSSESSIVE NOUNS

A plural possessive noun is a plural noun that shows ownership. You form most plural possessive nouns simply by adding an apostrophe (') after the final s of the plural noun, as in *students'*.

Not all plural nouns end in the letter s. Those that do not end in s are called *irregular* plurals. To form an **irregular plural possessive noun**, add an apostrophe and s ('s) to the end of the irregular plural noun.

For example:

Irregular Plural Noun	Irregular Plural Possessive Noun
men	men's
children	children's
mice	mice's
cacti	cacti's

Possessive forms can be tricky in writing, but using them correctly is important to make your meaning clear. It just takes practice.

A. Underline the correct irregular plural possessive noun in each sentence below.

1. We could tell by the hoof prints that we were in the (deers' / deer's) territory.
2. (Moose's / Mooses') growing antlers have a soft, furry covering.
3. (Men's / Mens') clothing is located on the second floor of the store.
4. Feta cheese is made from goats' milk or (sheeps' / sheep's) milk.
5. Every night, I read my favourite (childrens' / children's) story to my little brother.
6. (Bison's / Bisons') tails have a furry end called a tuft.
7. The (medias' / media's) influence on children is stronger than ever.

B. Change the underlined singular possessive noun to a plural possessive noun.

1. the <u>child's</u> books   the _____ books
2. the <u>woman's</u> group   the _____ group
3. the <u>goose's</u> habitat   the _____ habitat
4. the <u>ox's</u> yoke   the _____ yokes
5. the <u>person's</u> opinions   the _____ opinions
6. the <u>man's</u> shoes   the _____ shoes

C. Start a list or chart of irregular plural nouns and their possessive forms so that you can refer to it when you are writing.

LESSON 41: USE CONCRETE AND ABSTRACT NOUNS

You can group nouns into two categories: **concrete** and **abstract**.

Concrete nouns name people, places, or things that you can experience with your senses. A concrete noun is something or someone you can see or touch.
For example: cyclist, friend, prairie, gymnasium, notebook, windshield

Abstract nouns name ideas, feelings, or qualities. An abstract noun is something you cannot see or touch.
For example: courage, time, confusion, potential, safety, worry

When you are writing, you have a wealth of nouns to choose from. Always choose the best noun to describe exactly what you mean.

A. Write *C* for concrete or *A* for abstract on the lines beside the nouns.

1. weekend _____
2. mountain range _____
3. doubt _____
4. Lake Nipissing _____
5. batteries _____
6. luck _____

B. Underline the abstract noun in each sentence.

1. The nurse showed great kindness to our grandmother.
2. Natalie plays her violin with confidence on the stage.
3. Our classroom guest spoke to us about her experiences with bears.
4. This summer, many fields were flooded and crops were damaged.
5. My parents asked me to treat them with more respect.
6. Sidney shared his wisdom with the rest of the class.

C. Write two sentences containing abstract nouns. The sentences can also include concrete nouns.

1. _____
_____

2. _____
_____

D. Advertisers often use abstract nouns to influence people to buy things. Look for examples of abstract nouns in advertisements in magazines and online. Choose one example of an abstract noun in an ad and explain how its use is intended to affect consumers.

Grasp Grammar and Usage 75

LESSON 42: IDENTIFY ACTION, AUXILIARY, AND LINKING VERBS

There are several types of verbs. **Action verbs** show action, or something that a noun can do, such as *build* and *move*. They also show actions that cannot actually be seen, such as *wonder* and *learn*.

Some verbs have **auxiliary verbs** (helping verbs) before the main verb: *We are arriving. He does understand.* Common auxiliary verbs include forms of the verbs *be* (*am, is, are, was, were*), *do* (*do, does, did*), and *have* (*has, have, had*).

Linking verbs do not show action. They link the subject with more information about the subject: *She is relieved. The fabric feels soft.* Common linking verbs include *appear, seem, smell, sound, taste, look,* and forms of the verb *be.* Take care when identifying linking verbs, because some of them can also act as action verbs: *She feels the fabric. They look at the clock.*

Part of effective writing is choosing strong, exact, and informative verbs to make your meaning clear.

A. **Underline the action verb in each sentence.**

1. For his report, Trey researched the migration of Loyalists to Canada.
2. Many families in my community arrived from Holland in the 1950s.
3. Rose and her family participate in the Métis Community Festival every September.
4. Chantal found an online collection of documents from the War of 1812.
5. The Calvet family emigrated to Canada from France over 100 years ago.

B. **For each underlined verb in the sentences below, decide if it is an action verb, an auxiliary verb, or a linking verb. Write your choice on the line provided.**

1. The Talbot family had considered moving to Yukon. _____
2. Our lacrosse players appear surprised by the opposing team's speed. _____
3. My sister travelled to Venezuela as an exchange student four years ago. _____
4. We are creating a welcome sign for the school entrance. _____
5. The air smelled fresh after the spring rain. _____
6. The group disagreed on the issue of how to raise funds. _____
7. We did notice the deer's tracks during our trek through the woods. _____
8. The community was thrilled with the new fitness and biking trail. _____

C. **Write a verb to complete each sentence. For the action verbs, choose ones that are interesting and exact.**

1. On the worksite, the crane _____ the massive beams. (Use an action verb.)

2. We _____ driving to the airport when the car got a flat tire. (Use an auxiliary verb.)

3. As we _____ up the mountain, we had to stop to catch our breath. (Use an action verb.)

4. Throughout the day of competition, the skaters _____ confident. (Use a linking verb.)

5. They _____ organized a Community Clean-Up Day for the past eight years. (Use an auxiliary verb.)

6. The swimmer _____ through the choppy water of the lake. (Use an action verb.)

D. **Use one of the sentences from Exercise C as a starter for a short story. Use action, linking, and auxiliary verbs.**

LESSON 43
PROVIDE MORE INFORMATION: VERB PHRASES

A **verb phrase** consists of a main verb and one or more auxiliary (helping) verbs. It provides more information about the subject and the subject's action.

Some auxiliary verbs are forms of *be*, *do*, and *have*. An auxiliary verb is followed by a main verb.

For example: I *am searching* for my glasses.
(auxiliary verb = *am*; main verb = *searching*)
They *had watched* this movie many times.
(auxiliary verb = *had*; main verb = *watched*)
We *do hope* to win the trophy this year.
(auxiliary verb = *do*; main verb = *hope*)

A. **Underline the verb phrase in each sentence.**

1. She was stretching while she waited for her race.
2. For our schoolyard sale, we will gather clothing, books, and small toys.
3. They did learn about safety in science class, but they need reminders.
4. By January, Kazim had completed the first test in his swimming course.
5. When the fire alarm sounded, Cody's class was entering the gym.
6. I am experimenting with ways to train my dog to obey commands.
7. They have worked on the model village for two weeks, and it looks great!
8. I do enjoy visiting my cousin's farm every summer.
9. We should be practising our dance routine every day.

B. **Complete each sentence by writing a verb phrase.**

1. We _____ a reply by Wednesday.
2. The students _____ their presentations on Indigenous artists.
3. After the children _____ their dinner, they played until bedtime.
4. I _____ while I wait in the doctor's office.
5. We _____ soccer this Sunday afternoon.
6. They _____ about verbs last year, too.

LESSON 44

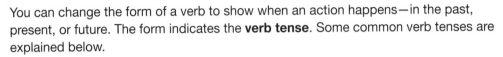
SHOW WHEN AN ACTION HAPPENS: VERB TENSES

You can change the form of a verb to show when an action happens—in the past, present, or future. The form indicates the **verb tense**. Some common verb tenses are explained below.

Simple present: the action is happening now, or it usually happens.
 For example: We *eat* dinner at 6 p.m.

Present progressive: the action is happening now and continues for a period of time.
 For example: I *am focusing* on my speaking skills this year.

Simple past: the action has already happened.
 For example: She *fixed* her bike on her own.

Past progressive: the action was ongoing in the past.
 For example: He *was shovelling* the sidewalk.

Simple future: the action has not yet happened.
 For example: They *will interview* two experts.

In your writing, use tenses to let your readers know when an action takes place. Maintain the same tense in a piece of writing unless you have a specific reason for changing it.

A. Identify the tenses of the underlined verbs in the sentences below.

1. We <u>explored</u> the wetland area along a boardwalk trail. _____

2. I <u>wonder</u> what role bears play in an ecosystem. _____

3. My friends <u>are monitoring</u> a dam. _____

4. Our class <u>will conduct</u> a study of our school's natural meadow area. _____

5. The scientist <u>was testing</u> the water for pollution. _____

B. For each sentence, underline the correct tense of the verb in parentheses.

1. In rainforests, funguses (grew / grow) on the forest floor and act as decomposers.

2. Last week, we (discovered / will discover) that someone had been dumping harmful substances in the river behind our house.

3. The more you research wetlands, the more you (will learn / learned) about how they act as filters.

4. When we (are walking / were walking) past the tide pool, my sister spotted some mollusks.

5. Today, bridges and tunnels called ecopassages (were helping / are helping) animals cross highways safely.

6. I (realize / will realize) that all living things are part of an ecosystem, but sometimes I question the contributions of mosquitoes!

C. **Identify the tense used in each sentence. Then, rewrite each sentence in two other tenses discussed in this lesson.**

1. I will assist with our school's EcoFair. _____

   a) _____

   b) _____

2. She is completing the final proofread of her poster. _____

   a) _____

   b) _____

D. **The following paragraph should have been written in the simple past tense. The first sentence is in the simple past, but all the other sentences have verb forms in other tenses. Rewrite this paragraph, replacing all of the underlined verbs with the past tense.**

Our class visited a freshwater marsh not far from our school to observe the wildlife. We <u>are seeing</u> a variety of birds, such as herons and ducks. We <u>were trying</u> to spot insects such as dragonflies and butterflies. Some plants we discovered <u>are</u> reeds and cattails. As for mammals, a muskrat or two <u>will make</u> an appearance.

_____

_____

_____

_____

_____

_____

_____

_____

_____

_____

_____

_____

E. **Identify the verb tense used in each sentence in Exercise D that has an underlined verb. How did you identify each tense?**

LESSON 45: MAKE THE PAST TENSE: IRREGULAR VERBS

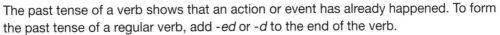

The past tense of a verb shows that an action or event has already happened. To form the past tense of a regular verb, add *-ed* or *-d* to the end of the verb.
   For example: offer – offered; receive – received

Some verbs, called **irregular verbs**, have a special spelling or use a completely different word for the past tense.
   For example: are – were; teach – taught; stand – stood

Some verbs are the same for both the present and the past.
   For example: read; quit; set

Knowing which verbs are irregular comes from writing and reading practice and from speaking and listening.

A. Underline the correct past-tense form for each verb.

1. shake – shaked / shook
2. ring – rang / ringed
3. hurt – hurted / hurt
4. ride – rided / rode
5. feel – feel / felt
6. wear – wore / wear
7. cut – cutted / cut
8. drink – drank / drinked

B. In each sentence, write the past tense of the verb in parentheses.

1. The raccoon got into our house because I did not _____ the back door. (shut)
2. We _____ we were going to win the game, but the other team scored two goals in the last five minutes of play. (think)
3. After playing in the tournament all day, I _____ for ten hours. (sleep)
4. My dad, who is a truck driver, _____ through two big storms yesterday. (drive)
5. I was disappointed, but I did not _____ our team's loss get me down. (let)

C. Write a sentence using one of the following verbs in the past tense: *feed*, *grow*, or *choose*.

_____
_____
_____
_____
_____

LESSON 46: USE PRESENT PERFECT AND PAST PERFECT TENSES

To express actions or events that have already taken place, we often use the simple past tense: *We walked*. However, there are also the **perfect tenses**: present perfect and past perfect.

The present perfect tense is used for actions of a non-specific time in the past or for actions that started in the past and are continuing into the present.

For example: We *have walked* for forty minutes, but we still haven't reached the lake.

This example is in the present perfect tense. It might sound odd to use the word *present* to express something that happened in the past, but it refers to the use of the present form of the verb *to have*.

The past perfect tense is used for actions that were completed sometime in the past before another action happened or to give a reason for something that happened.

For example: We *had walked* four kilometres before it started to rain.

To form the perfect tense of a verb, we use singular or plural forms of the verb *to have* plus the past participle of the verb.

A. Underline the verb in the perfect tense in each sentence. Identify it as either present perfect (*PrP*) or past perfect (*PP*).

1. I had completed three math problems before I realized I was on the wrong page. _____
2. When we arrived at the airport, I was happy to see that my grandparents' plane had landed. _____
3. We have lived in Alberta, Ontario, and British Columbia. _____
4. They had seen the movie six times, so they knew all the songs. _____
5. The students have constructed a triangle and a prism already. _____
6. The rabbit has eaten most of the cabbage we left out for it. _____

B. Underline the correct verb for each sentence.

1. Our lawn chairs (had flipped / has flipped) because of the strong winds.
2. Before we took a break we (had completed / had complete) three laps of the track.
3. She (have experienced / has experienced) many happy events in her life.
4. They (had seen / seen) many small towns on the coast before their trip was cut short.
5. The puppies (have grew / have grown) to be healthy, strong adult dogs.

C. Choose two sentences from this lesson and write an explanation of why the perfect tense is used in them.

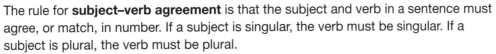

MATCH THE NUMBERS: SUBJECT–VERB AGREEMENT

The rule for **subject–verb agreement** is that the subject and verb in a sentence must agree, or match, in number. If a subject is singular, the verb must be singular. If a subject is plural, the verb must be plural.

For example: The *referee is* fair.
The *referees are* fair.

Some cases are trickier, such as the following:

Case	Explanation	Example
Collective noun	When the subject is a collective noun and is used as one unit, the verb is singular.	My *family likes* the museum. (one unit – singular verb)
Negative contractions	When the negative contractions *don't* and *doesn't* are used, the verb part of the contraction (*do, does*) agrees with the subject.	She doesn't worry. (singular subject *she* takes singular verb *does*) They don't worry. (plural subject *they* takes plural verb *do*)
Prounouns *that*, *which*, *who,* or *whom*	When the pronouns *that, which, who,* or *whom* are used, they take a singular verb if referring to a singular word and a plural verb if referring to a plural word.	The *book* that sits on this shelf *is* for our research. (*that* refers to the singular *book* and takes the singular verb *is*) The *books* that sit on this shelf *are* for our research. (*that* refers to the plural *books* and takes the plural verb *are*)

A. **Identify the underlined verb in each sentence as singular (S) or plural (P). Then, circle the subject that the verb agrees with.**

1. Amelie <u>doesn't</u> show any sign of nerves before her recital. _____
2. The firefighting team <u>moves</u> back from the blaze, unable to enter the building. _____
3. The boaters silently <u>waved</u> goodbye to their friends standing on the shore. _____
4. How <u>are</u> these stories similar? _____
5. Min-jun and Jakob <u>share</u> the same birthday. _____
6. The long-distance race <u>has started</u>, and it should be a close finish. _____

B. **Underline the correct verb in parentheses for each sentence.**

1. The chief and a band council member (is attending / are attending) the First Nations Environment Conference in Ontario.
2. The globe that (is / are) on the shelf will show us the latitude of Trois-Rivières.
3. Our group (is representing / are representing) our class at the assembly.
4. Standing by the entrance to the theatre, waiting for me, (was / were) my patient friends.

C. **Complete the following sentence starters. Use a verb that agrees with the subject. You may use any tense.**

1. Our school's environment club _____

2. The friends whom _____

3. Her mother and aunt _____

D. **The paragraph below has six errors in subject–verb agreement. Underline the incorrect verbs. Rewrite the paragraph by changing the verbs to a form that agrees with their subject. Keep the paragraph in the present tense.**

My friends Arun and Ivy and I am wondering what we could do today. Arun suggests that we play his drums and electric guitar, but apparently he doesn't understand the suffering that our "music" usually cause his family! A ride on the bike trails near our apartment building are the best suggestion so far. The morning is sunny and clear, perfect for riding. After our parents gives us permission, we gather our gear and pack lunches and water, and off we goes. Through wooded areas and open meadows our strong legs takes us. Arun grumble only twice about our choice to be outside.

_____

84   Grasp Grammar and Usage

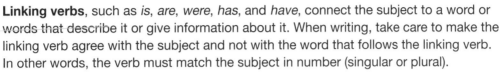

Linking verbs, such as *is*, *are*, *were*, *has*, and *have*, connect the subject to a word or words that describe it or give information about it. When writing, take care to make the linking verb agree with the subject and not with the word that follows the linking verb. In other words, the verb must match the subject in number (singular or plural).

    For example: Our favourite *game was marbles*.

                        Good *discussions have been* the best way to work out problems.

In the first example, the subject is *game*, which is singular, so the linking verb, *was*, must be singular. The linking verb does not have to agree with the plural noun *marbles*. In the second example, the subject is *discussions*, which is plural, so the linking verb, *have been*, must be plural. The linking verb does not have to agree with the singular noun *way*.

In your writing, remember that if your subject is singular, your linking verb must be singular. If your subject is plural, your linking verb must be plural.

A. In each sentence, underline the simple subject. (A simple subject is a single noun or pronoun.) Then, circle the correct linking verb.

1. One place I would like to visit again (is / are) Niagara Falls.
2. The pop-up advertisements on this website (was / were) a distraction.
3. Our backyard skating rink (has been / have been) the site of many neighbourhood parties.
4. Standing tall, the many flags (is / are) a symbol of our multiculturalism.
5. The new cycling and walking trail (has been / have been) a huge achievement for the town council.
6. Hurricanes and typhoons (was / were) the subject of my report.

B. Complete each sentence using one of the linking verbs below. The linking verb should agree with the subject. You can choose either past or present tense.

    is   are   was   were   has been   have been

1. The big bouquets of flowers _____ a special gift for our grandparents.
2. The response of the audience _____ an overwhelming surprise.
3. Canada's Eastern Arctic glaciers _____ an inspiring sight.
4. My aunts and uncles _____ an important part of my life.
5. A challenge for the pilots _____ the heavy crosswinds.
6. This vacation _____ the trip of a lifetime!

LESSON 49: REPLACE SUBJECT NOUNS: SUBJECT PRONOUNS

A pronoun is a word that takes the place of a noun in a sentence. The pronoun also replaces any other words that are associated with it, such as *the*, *and*, or *a*. We use pronouns to avoid repeating the same noun in related sentences.

>   For example: Our neighbour is a great person. Our neighbour helps us in many ways.
>   Our neighbour is a great person. She helps us in many ways.
>   (*She* takes the place of *Our neighbour*.)

A **subject pronoun** takes the place of the noun (or nouns) that is the subject of the sentence. It tells who or what the sentence is about, and it often appears at the beginning of a sentence. The subject pronouns include *I*, *you*, *she*, *he*, *it*, *we*, and *they*.

>   For example: Our new barns have been built. They are made out of steel.
>   (*They* takes the place of *Our new barns*.)

Note that *they* can be used as a gender-neutral singular pronoun. Other gender-neutral singular pronouns that have been used in recent years include *ze*, *xe*, and *ne*.

A. **Underline the subject pronoun(s) in each sentence.**

1. I am learning karate.
2. It was a great learning experience for all of us who visited the Woodland Cultural Centre.
3. You need to speak a little slower.
4. You and she have been selected as class representatives.
5. After careful thought, we decided to delay our trip to Newfoundland and Labrador until August.
6. After their graduation party, they groaned at the pile of dishes facing them.
7. He and I have become quite skilled in repairing small engines.
8. My cousin and I are auditioning for roles in the community play.

B. **On each line below, write a subject pronoun to replace the subject and any of its associated words in the first sentence.**

1. My new hearing aids arrive tomorrow. _____ fit in my ears perfectly.
2. Our school newsletter is published monthly. _____ is available online.
3. Macy and Sora are finalists in the speaking competition. _____ have one more round of speeches.
4. You and Darius are showing success in your work with the younger students. _____ are great role models.
5. He and I are starting a lawn-mowing business this summer. _____ already have some clients.

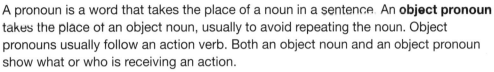

REPLACE OBJECT NOUNS: OBJECT PRONOUNS

A pronoun is a word that takes the place of a noun in a sentence. An **object pronoun** takes the place of an object noun, usually to avoid repeating the noun. Object pronouns usually follow an action verb. Both an object noun and an object pronoun show what or who is receiving an action.

For example: Avi lifted the box gently, and he set it on the shelf.

In this example, the object pronoun is *it*. *It* replaces *the box*. The object noun and object pronoun tell what Avi lifted.

Object pronouns can also follow a preposition, such as *to*, *from*, *with*, *for*, and *about*.

For example: Marta knows our Member of Parliament and often speaks to her.

In the example, *her* is the object pronoun, following the preposition *to*. *Her* replaces *our Member of Parliament*.

The object pronouns include *me*, *you*, *him*, *her*, *it*, *us*, and *them*.

Note that *them* can be used as a gender-neutral object pronoun. Other gender-neutral object pronouns that have been used in recent years include *zir*, *xem*, and *nem*.

A. **Read the first sentence in each set of dialogue. Underline the object noun and any associated words. Then, in the second sentence, write an object pronoun on the line provided to replace the words you underlined.**

1. "What is he doing with those straws?"   "He is building a model tower with _____."
2. "Please sing this song with me."   "I would, but I don't know _____."
3. "Can you tell Hannah to hurry?"   "I have not seen _____."
4. "Come swimming with Teo and me."   "I wish I could go with _____."

B. **Underline the object pronouns in the following sentences. There are other pronouns in the sentences, so choose carefully.**

1. I can't seem to control my dogs, so I will take them to obedience training.
2. We saw a documentary on the War of 1812, and I would like to learn more about it.
3. I miss my mom when she travels for work, so I always video-chat with her.
4. Our school sent Tomás to an ecology camp for a week, so we have made a Welcome Back sign for him for when he returns.
5. My older brother complains about my sister and me, but I think he will miss us when he goes to college.
6. Before our final game, I said to Katia, "The pressure doesn't even seem to worry you."

C. **Write a paragraph about what you did on the weekend, repeating one object noun in each sentence. Replace each repeated object noun in your writing with an object pronoun.**

LESSON 51
SHOW OWNERSHIP: POSSESSIVE PRONOUNS

A pronoun takes the place of a noun in a sentence. Like nouns, pronouns can show ownership—they can show that something belongs to someone or something. We call these pronouns **possessive pronouns**. They are never used with nouns or noun phrases. Possessive pronouns are used alone. They can be singular (*mine, yours, his, hers*) or plural (*ours, yours, theirs*).

For example: The painting of the horses is *mine*.
Is this cellphone *yours*?
I am happy with my decision, but she regrets *hers*.

You do not add an apostrophe to form these possessive pronouns.
Note that *theirs* can be used as a gender-neutral possessive pronoun. Other gender-neutral possessive pronouns that have been used in recent years include *zirs*, *xyrs*, and *nirs*.

A. For each of the personal pronouns, write its possessive form on the line beside it.

1. I   _____
2. you   _____
3. he   _____
4. she   _____
5. we   _____
6. you   _____
7. they   _____

B. Underline the correct possessive pronoun in each sentence.

1. I forgot my lunch, but I'm sure Teegan will let me share some of (hers / her's).
2. The guest speaker told all of us in the audience, "The responsibility is (yours / yours')."
3. Our neighbours' yard sale is on Saturday, and my family is having (ours / ours') on the same day.
4. Our Grade 1 reading buddies like to visit our classroom, but we usually go to (theirs / there's).
5. I am welcome to state my opinion, and you are welcome to state (your / yours).
6. My brother already cleaned his room, but I'll clean (mine / my) later.

C. Complete each sentence below with the possessive form of a personal pronoun.

1. I asked my friend, "Why did you take credit when you knew the idea was _____?"
2. Your team won this year, but next year the trophy will be _____!
3. This is the video Aneta produced; both the script and the music are _____.

LESSON 52: USE INDEFINITE PRONOUNS

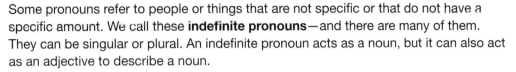

Some pronouns refer to people or things that are not specific or that do not have a specific amount. We call these **indefinite pronouns**—and there are many of them. They can be singular or plural. An indefinite pronoun acts as a noun, but it can also act as an adjective to describe a noun.

Singular indefinite pronouns include *another*, *each*, *everything*, *somebody*, *something*, *other*, and *much*.
  For example: This piece of cake makes me want to have *another*.

Plural indefinite pronouns include *others*, *several*, *many*, *few*, and *both*.
  For example: *Few* can say that they have been in outer space.

Make sure that the indefinite pronoun and the verb agree—a singular pronoun takes a singular verb and a plural pronoun takes a plural verb.

A. Underline the indefinite pronoun in each sentence. Then, circle the verb in parentheses that agrees with it.

1. Ayesha discovered that something (was / were) wrong with the results of her experiment.
2. This website is fun, but the other (seem / seems) more informative.
3. Several mechanics (has / have) tried to repair the lawn mower, but it's still broken!
4. My dad always smiles and says, "Everything (is / are) possible."
5. Youth from all over British Columbia attended the gathering, where much (is / are) taught by the Elders.
6. A lot of people try to scale the wall, but many (fall / falls) on their first try.

B. Place each of the indefinite pronouns in the sentence where it makes the most sense. Write them on the lines provided.

much   both   each   somebody   others

1. I detect one problem with their model plane's design. Can you detect any _____?
2. Darnell prefers history, and Morgan likes art, but I like _____.
3. Is there _____ at the door?
4. _____ has been written about Canada's Arctic, but I would like to experience it for myself.
5. I researched various perspectives on the new housing development, and _____ is slightly different.

C. Skim a book to search for indefinite pronouns that act as nouns. Write three sample sentences and underline the indefinite pronoun in each one. Then, write a sentence using one of the indefinite pronouns.

LESSON 53: USE REFLEXIVE PRONOUNS

One type of pronoun is the **reflexive pronoun**. We use a reflexive pronoun when the pronoun, which is also an object pronoun, reflects back to the subject of the sentence. In other words, we use a reflexive pronoun in the case of something doing something to itself.

　　For example:　The friends are treating *themselves* to ice cream after their game.

In the example, *themselves* is the object of the verb *treating*, and it reflects the subject, *friends*. Without the reflexive pronoun, there would be repetition: The friends are treating the friends to ice cream after their game.

Reflexive pronouns are both singular—*myself, yourself, himself, herself,* and *itself*—and plural—*ourselves, yourselves,* and *themselves*.

A. Write the reflexive pronoun that refers back to the subject in each short sentence.

1. I will help _____.
2. They can teach _____.
3. We won't hurt _____.
4. You students can introduce _____.
5. Paula, don't blame _____.
6. He trusts _____.
7. The computer's calendar resets _____.
8. She knows _____.

B. Write the reflexive pronoun form of the pronoun in parentheses.

1. We excused _____ (we) and left the table.
2. I promised _____ (me) that I would be more positive about my accomplishments.
3. My fellow citizens, you should ask _____ (you) how much you care about our community's future.
4. The Grade 4 students congratulated _____ (them) on making all the paintings for the hallway.
5. Mrs. Sedek said, "Alisha, you are not acting like _____ (you) today."
6. The bird is saving _____ (it) from harm by staying in the cedars.

C. Write a sentence for three of the reflexive pronouns found in this lesson.

1. _____

2. _____

3. _____

D. Write a descriptive paragraph about a class trip, such as a trip to a conservation area. Use at least three different reflexive pronouns.

LESSON 54
MAKE PRONOUNS AND ANTECEDENTS AGREE

The noun or noun phrase that a pronoun takes the place of is called an **antecedent**, meaning *something that comes before something else*. A pronoun must always agree with its antecedent. If the antecedent is singular, the pronoun must be singular. If the antecedent is plural, the pronoun must be plural. If the antecedent is female or male, the pronoun must agree in gender.

For example: I told Natalie I would join her for lunch.
(the noun *Natalie* is the antecedent of the pronoun *her*)
The eaglets are in the nest, but I can't see them.
(the noun *eaglets* is the antecedent of the pronoun *them*)
My oldest brother says the sculpture is his.
(the noun phrase *oldest brother* is the antecedent of the pronoun *his*)

A. **In each sentence, circle the pronoun that replaces the underlined antecedent.**

1. The Nahanni River is a World Heritage Site because it contains so many unique geological features.

2. Many people live in tropical rainforests, where they have adapted to the environment.

3. My friend lives on the north shore of Lake Superior, where he helps with the family business.

4. My partner and I tried, but we could not find climate data for that area of Thailand.

5. Volcanoes are fascinating, but I wonder what living near them is like.

6. Luke gasped when he dropped the new cellphone on the sidewalk.

7. Mo was thrilled to learn that the winning ticket was his.

B. **Underline the pronoun that has an antecedent in each sentence. Circle the antecedent.**

1. When Madison was sick for a long time, our class sent her a card.

2. I saw the candle flickering dangerously high and raced to put it out.

3. Erik and Pavani are co-writers of the play they are presenting.

4. By the time Julia and I arrived at skating practice, we had missed the warm-up.

5. Although the admission was difficult, Jon knew the fault was his.

6. The girls made an inspiring video and submitted it to a contest.

7. The phone stopped ringing as soon as I found it.

C. **Look through some writing you have done to identify any pronouns that do not agree with their antecedents and make the necessary corrections. Create a reminder for yourself about the rule for pronoun–antecedent agreement.**

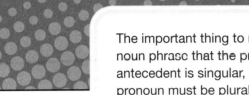

A VARIETY OF PRONOUNS AND ANTECEDENTS

The important thing to remember about **pronouns** and **antecedents** (the noun or noun phrase that the pronoun takes the place of) is that they must always agree. If the antecedent is singular, the pronoun must be singular. If the antecedent is plural, the pronoun must be plural. If the antecedent is masculine or feminine, the pronoun must agree in gender. The rule is true for all types of pronouns, including subject, object, demonstrative, and reflexive.

For example: *The boy* was proud of *himself*.
  The lions surrounded the gazelles *they* were hunting.

A. **In each sentence, circle the pronoun that replaces the underlined antecedent.**

1. Please ask <u>the new player</u> if (they / she) has the proper safety equipment.
2. <u>The wayward hikers</u> found (themselves / itself) in a beautiful mountain meadow.
3. When I take <u>my little sister</u> to the park, I always have fun playing with (them / her).
4. <u>Kindness and hope</u> are qualities I admire, and I strive to achieve (it / them).
5. Next week, <u>Janie and I</u> are in a dance performance, and you should come see (us / her).
6. <u>Liang</u> would like to attend camp, but (he / they) might have to go to summer school instead.
7. If <u>the recycling bin</u> is full, you should empty (it / them).

B. **Each of the following sentences has a pronoun that does not agree with its antecedent. Write a corrected version of the sentences.**

1. Mary Pratt is a Canadian painter, and I intend to do some research about them.

   _____
   _____

2. We have more items to add to our work safety display, but we don't really need it.

   _____
   _____

3. Mikayla and I could not stop myself from laughing at the cat videos.

   _____
   _____

C. **Skim some pages from a magazine, comic book, or book to find examples of sentences with pronouns. Choose five sample sentences, identify the pronoun and antecedent, and explain how they agree.**

LESSON 56
WRITE DESCRIPTIVE WORDS: ADJECTIVES

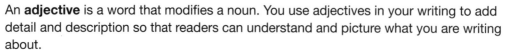

An **adjective** is a word that modifies a noun. You use adjectives in your writing to add detail and description so that readers can understand and picture what you are writing about.

    For example: We entered the *spacious*, *orderly* rec room.

There are different types of adjectives. Some of them specify or limit the noun they modify.

    For example: *daycare* worker, *government* service (nouns as adjectives)
                that column, *this* author (pronouns as demonstrative adjectives)
                her understanding, *their* townhouse (pronouns as possessive adjectives)

Adjectives make writing interesting, descriptive, and precise. Select the best adjectives for your purpose and use a variety of them.

A. Underline the adjectives in each sentence. The number in parentheses tells you how many adjectives to find. Do not include articles such as *the* and *a*.

1. The restless crowd waited impatiently in the busy airport lounge. (3)
2. A mature Pacific salmon will travel many kilometres to reach its original home. (5)
3. In this portrait, the king has a kind, laughing expression, and his crown sits at a peculiar angle. (5)
4. As the terrified boy walked down the long hallway to his room, he realized he should not have watched that second horror movie. (6)
5. That display of modern photographs explains how resources are used in Inuit culture. (3)
6. The mighty waves of the angry ocean almost knocked the young woman off her feet as she stood on its shore. (5)

B. Add two adjectives to modify the underlined noun in each sentence. Choose adjectives that allow a reader to imagine what you describe.

1. I polished the exterior of the _____, _____ truck.
2. He added the vegetable scraps to the _____, _____ compost bin.
3. When she broke her arm, she felt a _____, _____ pain.
4. After a long day of ice fishing, they were glad to get back to the _____, _____ cabin.
5. She couldn't stop staring at her _____, _____ lizard.
6. In autumn, many animals migrate away from the _____, _____ tundra.

94   Grasp Grammar and Usage

C. For each underlined adjective, think of three more descriptive and precise adjectives to describe the noun.

1. a <u>good</u> song _____ _____ _____

2. a <u>little</u> animal _____ _____ _____

3. a <u>happy</u> person _____ _____ _____

4. a <u>nice</u> day _____ _____ _____

D. Write a descriptive paragraph about your home or neighbourhood. Use a variety of descriptive adjectives and adjectives that help give specific information, such as demonstrative adjectives and possessive adjectives.

_____
_____
_____
_____
_____
_____
_____
_____
_____
_____
_____
_____
_____
_____
_____
_____
_____
_____

E. Create a two-column chart to record adjectives you tend to overuse in your writing and adjectives you can use to replace them. Skim advertisements and pages from magazines and books, or consult a thesaurus for ideas.

MAKE COMPARISONS: ADJECTIVES

An **adjective** has two different forms for making comparisons: comparative (for comparing two things) and superlative (for comparing three or more things). The chart below shows how to form different types of adjectives into adjectives that make comparisons.

Types of Adjectives	Positive	Comparative	Superlative
regular, one syllable	light	lighter	lightest
one syllable ending in a consonant with a single vowel before it	mad	madder	maddest
two syllables ending in a consonant that is followed by the letter *y*	silly	sillier	silliest
two or more syllables, other cases	anxious	more anxious	most anxious

A. For each adjective, write the comparative and superlative adjective forms.

                Comparative   Superlative

1. beautiful _____ _____
2. amazing _____ _____
3. healthy _____ _____
4. bright _____ _____
5. guilty _____ _____

B. Complete the sentences below. On each line, write the correct form of the adjective in parentheses.

1. The car you polished is _____ than mine. (shiny)
2. The person who cheated should feel the _____ of all. (disappointed)
3. Of the two of us, I feel that I am _____. (responsible)
4. This blue line appears to be _____ than the green one. (thin)
5. I calculated all the distances, and this route will be the _____. (quick)
6. This summer was the _____ summer ever recorded. (hot)
7. This weather makes our cat _____ than usual. (lazy)

C. Write a poem or song using the positive, comparative, and superlative forms of one or more adjectives.

LESSON 58: DESCRIBE ACTIONS: ADVERBS

An **adverb** is a word that modifies a verb, adjective, or another adverb. When modifying a verb, an adverb tells how, when, where, how often, or to what degree the verb's action is done. Many adverbs end in -*ly*, but there are also many adverbs that do not.

For example: They left *quietly*. They walked *outside*.

When modifying an adjective or adverb, an adverb makes the meaning more precise.

For example: I entered the *surprisingly* quiet room. (adverb modifying the adjective *quiet*)

He can jump *extremely* high. (adverb modifying the adverb *high*)

In your writing, use a variety of adverbs to provide specific information about the actions you describe.

A. For each adverb in italics, underline the word that it modifies. Then write "Verb," "Adjective," or "Adverb" to state what part of speech the modified word is.

1. He contributed *generously* to the school's fundraiser. _____
2. Our arrival resulted in two *very* happy grandparents. _____
3. We *eventually* solved the math problem and proceeded to the next one. _____
4. The evening stars gleamed *astonishingly* brightly. _____
5. British Columbia has *majestically* tall redwood trees. _____
6. She presented her report *extremely* confidently. _____
7. The yogurt was *shockingly* high in sugar. _____

B. Underline the adverbs in the sentences below. The number in parentheses tells you how many adverbs there are.

1. Shay visits her cousin quite regularly, and they always have fun. (3)
2. My grandmother frequently speaks Cree, and I am extremely proud because I am learning it fairly quickly. (4)
3. We conducted our science experiment very safely. (2)
4. The incredibly lucky boys won passes for their favourite band's concert, and they were too excited to speak! (2)

C. Write two sentences. Include at least one adverb in each sentence.

_____

_____

MAKE COMPARISONS: ADVERBS

An **adverb** has two different forms for making comparisons: comparative and superlative. The chart below shows how to change the positive (regular) form of different types of adverbs into adverbs that make comparisons.

Types of Adverbs	Positive	Comparative	Superlative
one syllable; adverbs that have the same form as the adjective	fast	faster	fastest
two or more syllables ending in *-ly*	politely	more politely	most politely
irregular	little	less	least
	much	more	most

A. On each line, write the correct form of the adverb that is in parentheses.

1. She has been waiting _____ than I have. (long)

2. Jelena threw the ball _____ on the last of three tries. (confidently)

3. Of all his teammates, Avi cares the _____ about winning every game. (little)

4. Melissa reacted _____ than her brother to the news. (positively)

5. The performances showed that the winning band had practised the _____ of all the competitors. (much)

6. We could conduct this experiment _____ by wearing gloves. (safely)

7. Diane travelled the _____ to get to my party. (far)

B. Write two sentences, one with a comparative adverb and one with a superlative adverb. You can use the words provided below, select words from the list above, or think of your own words. Be sure that you use the word as an adverb and not as an adjective.

cheerfully   quietly

_____

_____

_____

C. Choose three animals and write a paragraph comparing how they move and act. Use at least three comparative and three superlative adverbs in your comparison.

LESSON 60: WRITE DESCRIPTIVELY: ADJECTIVES AND ADVERBS

Adjectives describe nouns and pronouns. **Adverbs** describe verbs, adjectives, and other adverbs. Knowing which words are adverbs and which words are adjectives is important, but identifying them can be tricky.

Adverbs and adjectives often look similar. If you are not sure whether a word is an adjective or an adverb, look at what the word is doing or describing. You can also look at what questions the word is answering.

Adverbs often answer the questions How? When? Where? To what extent? and Why? Adjectives often answer the questions What kind? Which? and How many?

Sometimes, the same word can be either an adjective or an adverb, such as *fast*, *late*, or *straight*.

In your writing, use a variety of adjectives and adverbs to provide accurate, interesting descriptions.

A. For each sentence, decide whether the underlined word is an adjective or adverb. Check the appropriate box.

1. She is behaving <u>unreasonably</u>. ❏ Adjective ❏ Adverb
2. The decision has <u>global</u> effects. ❏ Adjective ❏ Adverb
3. They rushed <u>straight</u> to the ticket window. ❏ Adjective ❏ Adverb
4. He was breathing <u>hard</u> after his goal. ❏ Adjective ❏ Adverb
5. They knew the distance was <u>far</u>. ❏ Adjective ❏ Adverb
6. Let's go to the <u>early</u> show. ❏ Adjective ❏ Adverb

B. Use adjectives and adverbs to make the sentences below more descriptive. Write the revised sentences on the lines provided.

1. The boat moved on the lake.

   _____

2. The toddler caught the ball.

   _____

3. The person played the instrument.

   _____

C. Revisit some descriptive writing you have done. Consider how adding adverbs and adjectives, or selecting more precise ones, could improve its accuracy and interest. Select three sentences to rewrite.

SHOW RELATIONSHIPS: PREPOSITIONS

A **preposition** is a word that links a noun or pronoun to another word or part of a sentence. Prepositions help to show the time, place, or the manner in which something happens or to indicate some other relationship between the noun and the other word or part of the sentence.

For example: The audience laughed *throughout* the speech.

In the example, the preposition *throughout* connects the noun *speech* with the verb *laughed*.

Some common prepositions are *up*, *down*, *at*, *to*, *on*, *in*, *for*, *with*, and *from*. Some other examples are *over*, *by*, *near*, *after*, *before*, *until*, *since*, *across*, *within*, *unlike*, *except*, and *following*.

A. **Underline the prepositions in each sentence. The number in parentheses tells you how many there are.**

   1. For many years, my family has gone to a Remembrance Day ceremony in our community. (3)

   2. On Monday, our class went to a Indigenous art gallery and toured it with a guide who explained themes and symbolism in the works. (4)

   3. In Morden, Manitoba, near the border with the United States, you can dig for fossils at the Canadian Fossil Discovery Centre, after you "meet" Bruce and Suzy, two giant mosasaur fossils. (6)

   4. My aunt has travelled across Canada from the Atlantic Ocean to the Pacific and to its northern territories; she said that the northern lights is a spectacle unlike any other. (5)

B. **Complete each sentence with one of the prepositions listed below. Use a different preposition in each sentence. Not all of the prepositions will be used.**

   down   except   by   until   since   within   over   following

   1. _____ the game, the team went out for burgers.

   2. She had never really appreciated people's kindness _____ yesterday, when she needed help.

   3. _____ two days of leaving for camp, Matt had become sick with the flu.

   4. He noticed an increase in his flexibility _____ he started doing yoga.

   5. Everyone had tried tai chi before, _____ Justine.

   6. We had to move our club meeting to a room _____ the hall.

100   Grasp Grammar and Usage

LESSON 62: RECOGNIZE PHRASES: PREPOSITIONAL PHRASES

A **prepositional phrase** is a group of words that begins with a preposition and contains a noun or pronoun. That noun or pronoun is the *object of the preposition*. The object in the phrase links to the preposition, and the phrase itself links to the rest of the sentence.

> For example: The sand blew *across the dunes*.

In the example, the preposition *across* links to the object *dunes*; the prepositional phrase *across the dunes* tells where the sand blew.

Prepositional phrases can act as adverbs or as adjectives. They help to answer the questions Where? How? When? Why? To what extent? or Which one?

> For example: He ended the routine *with a back flip*. (tells how)
> They will do homework *after the game*. (tells when)
> She was exhausted *from the triathlon*. (tells why)
> The boy *in the video* is my friend. (tells which one)

A. Underline the prepositional phrases in each sentence. The number in parentheses tells you how many there are. Then, circle the preposition in each phrase.

1. The concert ended with three standing ovations, and it was 11 p.m. before we left. (2)
2. Over three months, the youth group held five fundraisers for their charities. (2)
3. My family and I strolled through the park before stopping at a restaurant for falafels. (3)
4. In the souvenir shop, I chose the black hat with the maple leaf. (2)
5. In the fall, Maddie will help with the harvest on her parents' farm. (3)
6. At the bowling alley, Kip bowled three strikes in a row. (2)
7. The woman in the green coat has been waiting by the door for thirty minutes. (3)

B. Complete each of the sentence starters by adding two prepositional phrases.

1. She walked _____

2. I felt tired _____

3. They described their trip _____

C. Write four sentences that each contain a prepositional phrase for a different purpose—to describe how, when, why, and which one. Underline each prepositional phrase and identify its purpose.

LESSON 63

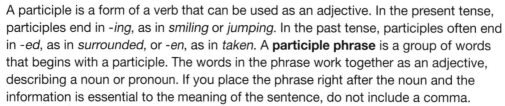
DESCRIBE A NOUN: PARTICIPLE PHRASES

A participle is a form of a verb that can be used as an adjective. In the present tense, participles end in *-ing*, as in *smiling* or *jumping*. In the past tense, participles often end in *-ed*, as in *surrounded*, or *-en*, as in *taken*. A **participle phrase** is a group of words that begins with a participle. The words in the phrase work together as an adjective, describing a noun or pronoun. If you place the phrase right after the noun and the information is essential to the meaning of the sentence, do not include a comma.

For example: The team *wearing the blue shirts* is playing well. (describes the noun *team*)

You can also place participle phrases at the beginning of sentences. Placing them at the beginning sometimes can help you vary your sentence structures. If you place the phrase at the beginning of a sentence, follow it with a comma.

For example: *Covered in mud*, the girls just grinned. (describes the noun *girls*)

A. **In each sentence, underline the participle phrase. Then, circle the noun or pronoun that it describes.**

1. Clutching the railing, we ventured onto the walkway over the canyon in Capilano Park.
2. The big dog waiting patiently by the gate belongs to me.
3. Encouraged by the applause, the singer began to relax.
4. The apples fallen on the ground around the trees were still good to eat.
5. Frozen in alert stillness, the deer sensed a predator.
6. Reading about her ancestors, Alana felt a strong need to learn to speak German.
7. The interview posted online provides Mathieu with valuable first-hand information.
8. The girl on the dock studying the stars might be an astronomer someday.
9. The children were happy with the stories chosen by the librarian for story hour.

B. **Use each participle phrase below in a sentence. You may place it at the beginning of the sentence or elsewhere. Use punctuation correctly.**

1. consuming the berries

   _____

2. chosen by her teammates

   _____

C. **Choose a sentence from Exercise A and write another sentence to follow it. Use a participle phrase in your sentence.**

LESSON 64: BE CLEAR: MISPLACED MODIFIERS

A modifier is a word or phrase that modifies (or describes) the meaning of something else in a sentence. Adjectives and adverbs are modifiers. When you use modifiers, it is very important to place them correctly in the sentence or your meaning may not be clear. To avoid **misplaced modifiers**, keep them as close as possible to what they are modifying, or move them to a place in the sentence that will make your meaning clear.

For example: The *hissing* boy's cat was scary. (misplaced modifier)
The boy's *hissing* cat was scary. (clear modifier)

The first sentence makes it seem that the boy, not the cat, is hissing.

Be aware that the placement of some modifiers can result in very different meanings.

For example: She *almost* won every award. (*almost* modifies *won*)
She won *almost* every award. (*almost* modifies *every award*)

The first sentence above means that she came close but did not win any awards. The second sentence means that she won most of the awards. The placement of *almost* completely changes the meaning.

A. **The underlined words below are misplaced modifiers. Rewrite each sentence, placing the modifier in a position that makes more sense.**

1. My <u>plastic</u> sister's toys are scattered on the playroom floor.

   _____

2. They walked onto the stage to present the play they wrote <u>nervously</u>.

   _____

3. The boy who was wakeboarding <u>expertly</u> waved to his friends on shore.

   _____

4. The <u>antique</u> farmer's tractor was next on the auctioneer's list.

   _____

B. **Read this sentence: *I lost all the fundraising money*. Then make two sentences with very different meanings by placing the modifier *almost* in different places in each sentence.**

1. _____

2. _____

C. **Find at least two examples of modifiers in sentences and try placing them in a different spot in the sentences. Write an explanation of how sentence clarity is affected when the modifier is moved.**

BE CLEAR: MISPLACED AND DANGLING MODIFIERS

A **modifier** is a word or phrase that modifies the meaning of, or describes, something else in a sentence. If modifying phrases are misplaced in a sentence, confusion—sometimes funny confusion—may result.

For example: *Stuffed inside the backpack*, my brother found his pen.

The brother is not stuffed inside the backpack! To fix a misplaced modifying phrase, place it as close as possible to the word it modifies—in this case, *pen*.

For example: My brother found his pen *stuffed inside the backpack*.

A related problem is the dangling modifier. A modifier dangles when it is meant for a word that is not actually in the sentence. Instead, it modifies a word that it was not intended for.

For example: *Walking along the beach*, a whale suddenly breached.

The whale is not walking along the beach! To fix a dangling modifier, include the word it modifies—in this case, *Mom*. Also, add any other words that are needed.

For example: *As Mom* was walking along the beach, a whale suddenly breached.

A. **The following sentences have misplaced modifiers. Write the correct version of each sentence on the lines below.**

1. The zookeeper fed the lions wearing an orange vest.

2. The lions pounced on the food lounging in the shade of a tree.

B. **The following sentences have dangling modifiers. Write a correct version of each sentence. Keep the modifier (or modifying phrase) the same, but you can change or add other words and change the order of words.**

1. Playing in the park, the sun felt wonderful.

2. Covered in glitter and glue, the craft time came to an end.

3. After jumping on the trampoline, the horses needed to be fed.

LESSON 66: USE JOINING WORDS: CONJUNCTIONS

A **conjunction** is a word that joins elements such as words, phrases, or clauses in a sentence. Coordinating conjunctions such as *and*, *or*, *but*, *so*, *yet*, and *nor* are used to connect similar elements in a sentence.

For example: We swept the floor *and* cleaned the windows.

Subordinating conjunctions join a subordinate clause to a main clause. The clause introduced by the subordinating conjunction is dependent on the main clause; it tells something about the main clause but cannot stand on its own. Examples are *after*, *before*, *because*, *unless*, *if*, *when*, *where*, *so that*, *since*, and *while*.

For example: I am crying *because* I am chopping onions.

You can place a subordinate clause before or after the main clause. If you put it before the main clause, always follow it with a comma.

A. In each sentence, underline the conjunction. On the line that follows the sentence, identify the type of conjunction by writing *C* for coordinating conjunction or *S* for subordinating conjunction.

1. I hope my sister will help me with my project since I helped her with hers. _____
2. Calvin wants to join the birders' group, so he will need to borrow binoculars. _____
3. I hope to study welding and plumbing in college. _____
4. Unless I find more sources, I won't have enough information for my report. _____

B. In each sentence, underline the correct conjunction from the choices in parentheses.

1. Tal shot the video of the historian (so that / while) I asked the questions.
2. (After / So) showing leadership on the field trip, Amy was elected class representative.
3. In science class, we will explore heat loss (when / yet) we get two more samples of insulation.
4. (Unless / Since) unexpected clouds move in, we will have good conditions for observing the stars.
5. I store my science instruments (for / so) my little brother cannot find them.
6. My computer has been much faster (since / when) my sister installed more memory.

C. Complete each sentence starter by adding a conjunction and a related idea.

1. I promise I will visit you this summer _____.
2. Our group will succeed _____.

D. Write one sentence that has a coordinating conjunction and one sentence that has a subordinating conjunction. For each sentence, describe the role of the conjunction.

LESSON 67: EXPRESS EMOTIONS: INTERJECTIONS

Writers sometimes insert a word or short phrase in their writing to express a feeling or emotion such as surprise, disgust, excitement, disbelief, or agreement. These words are called **interjections**. You can use interjections in dialogue or in regular text. If the interjection shows strong emotion, place an exclamation mark after it. If the emotion is mild, use a comma.

For example: *Ouch!* Your words hurt more than a bee sting.
Hmm, I will have to give your idea some thought.

Other examples of interjections are *bravo, yuck, phew, ha, hey, yikes, yes, oh, no,* and *well*.

Use interjections to convey emotion, create voice, and give your writing some liveliness and variety—but be careful not to overuse them. Use them more in informal writing than in formal writing.

A. For each sentence, choose which of the two interjections in parentheses makes more sense. Write the interjection and add the punctuation that you think should appear after it.

1. _____ I still have six more cartons to carry. (Phew / Yikes)
2. _____ I completely forgot to tell my parents about the school bake sale! (Oh no / No)
3. _____ despite what some people say, I think our pancakes tasted fine. (Well / Wait)
4. _____ it's true—we do have a math test today. (Yes / Phew)
5. _____ Your drum solo was amazing! (Yikes / Bravo)
6. _____ I wonder what happened. (Hmm / Ouch)
7. _____ I knew I could fix that computer! (No / Ha)

B. Choose four interjections from Exercise A, or choose other interjections that you know, and use each one in a sentence.

1. _____
2. _____
3. _____
4. _____

SECTION REVIEW

A. **For each noun, identify its type.**

	Common, Collective, or Proper?	Concrete or Abstract?
1. team	_____	_____
2. belief	_____	_____
3. Cape Breton Island	_____	_____
4. fossil	_____	_____

B. **Identify the type of verb in italics in each sentence.**

1. We *are* constructing a compost bin.   action   auxiliary   linking
2. The class *tested* the properties of three substances.   action   auxiliary   linking
3. The audience *is* uncomfortable in the sticky heat.   action   auxiliary   linking
4. That architect *designs* homes for senior citizens.   action   auxiliary   linking

C. **For each sentence, identify the tense of the verb in italics as present, present progressive, past, or future.**

1. I *concentrated* during the math test and was successful. _____
2. We *are estimating* the amount of wood needed for our model tower. _____
3. Dragonflies are strong fliers and easily *change* direction. _____
4. My sister *will graduate* from college this spring. _____

D. **For each sentence, underline the simple subject. Then, circle the verb that shows correct subject–verb agreement.**

1. Badminton (is / are) a favourite sport of mine.
2. The memorial plaques (honours / honour) the city's war veterans.
3. She (don't / doesn't) care about having fashionable clothes.
4. The committee (acts / act) as our voice.
5. The items that are in the top cupboard (are / is) for after-school snacks.

E. **Each of the following sentences has a pronoun that does not agree with its antecedent. Write a corrected version of each sentence.**

1. Petra and Jackson helped ourselves to the fruit.

   _____

2. We saw a play last night and really enjoyed them.

   _____

F. **Underline the adjectives in each sentence. Do not include articles such as *the* and *a*. The number in parentheses tells you how many adjectives are in each sentence.**

1. The three girls have been great friends since they met in the park playground. (3)
2. The cheetah is the fastest land mammal and lives in the tall, dry grass of African plains. (5)
3. I tried that tricky soccer move yesterday, but I need more coaching. (3)
4. My favourite Canadian holiday is July 1 because our town has a fun, multicultural parade. (4)

G. **For each adverb in italics, underline the word that it modifies. Then identify whether the word is a verb, adjective, or adverb.**

1. A peregrine falcon can attack its prey *incredibly* quickly. _____
2. We *patiently* waited for the food delivery. _____
3. We had some *very* surprising temperatures last spring. _____
4. He wobbled *slightly* in his skating routine. _____

H. **Underline the prepositions in each sentence. The number in parentheses tells you how many prepositions are in each sentence.**

1. Until this year, I didn't pay much attention to our town's history. (2)
2. The butterfly garden on our school grounds was planted before I started Kindergarten. (2)
3. Following the introduction, the speaker walked across the stage and stood beside the screen. (3)
4. The goats usually stay within their boundary, but sometimes they get over the fence in one amazing leap. (3)

I. **The underlined words are misplaced modifiers. Rewrite each sentence, placing the modifier in a position that makes more sense.**

1. The sculptor who had been working <u>carefully</u> dropped the sculpture on his foot.

   _____
   _____

2. The <u>delicious</u> family's dinner was prepared by the children.

   _____
   _____

3. She tripped over the <u>purple</u> child's shoe.

   _____
   _____

J. **In each sentence, underline the modifying phrase and circle the word that it modifies.**

1. Guided by our expert, we set off in our kayaks.
2. That woman standing by the door is a Canadian opera singer.
3. Meg, confused by the instructions, asked the teacher for help.

K. **Write two paragraphs describing the physical landforms and climate of your community. In your writing, use at least two coordinating conjunctions (*and, or, but, so, yet, nor*) and at least two subordinating conjunctions (*after, before, when, while, because, so that, since, where, if, unless*). Circle all the conjunctions.**

L. **Write two descriptive paragraphs about something you found challenging but eventually succeeded in doing. In your writing, include effective and accurate adjectives and adverbs that will help readers picture what you are describing. Also, be sure that you have placed the modifiers correctly.**

CRAFT AND COMPOSE

Like math, writing has rules that need to be followed in a logical way. Like painting, writing also has methods that can be used to create different effects in a creative way.

Good writers craft each word, sentence, paragraph, or line of poetry with an art that makes their readers want to read more. This doesn't happen by accident: it comes from thinking about *what* you want to write, *why*, and for *whom*—and then *how*. Once you identify your topic, purpose, and audience, you can make the right decisions about how to grab your readers and get your message across to them clearly.

In this section, you will learn how your choices as a writer can make your writing lively and powerful.

"The English language is a work in progress. Have fun with it."

— Jonathan Culver

LESSON 68: CREATE A LIFE MAP: CHOOSING A TOPIC

Writing can be extremely rewarding. **Choosing a topic** or an idea that interests you or that you feel strongly about can make the task of writing enjoyable.

Sometimes, it helps to "write what you know" and write about an important or memorable event in your life. Creating a life map is a good way to come up with ideas to use in your writing. A life map is a graphic organizer in which you can record a series of important and meaningful events that have happened in your life.

Here are some examples of things to include in your life map:

- your first memory
- joining a team or club
- a time you hurt yourself or got sick
- a vacation
- an important achievement

These are just a few ideas; there are many more options. Include anything from your life that is memorable and that you feel strongly about. Once you have created your life map, choose an event from your life to write about.

A. Complete the life map below. Write important events from your life in each oval, in the order that they occurred.

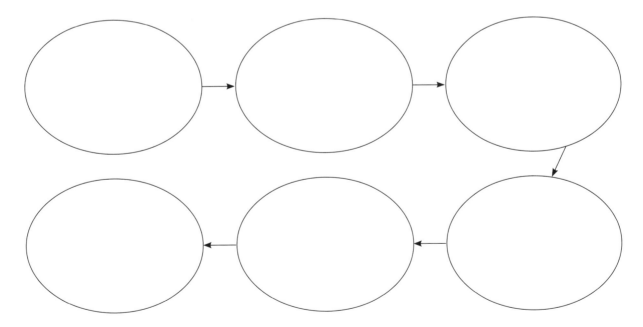

B. Choose one event from your life map to write about. Write a short descriptive paragraph about the event. Pick an event that you feel strongly about.

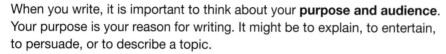

When you write, it is important to think about your **purpose and audience**. Your purpose is your reason for writing. It might be to explain, to entertain, to persuade, or to describe a topic.

Your audience is any person or group who will read your work. It might include your classmates, friends, teacher, family, people in the community, or even yourself.

Before writing, it is important to think about your purpose and audience. This will help you choose the right voice based on who will be reading your work.

For example, if you are writing a letter to your teacher, you will use a more formal voice. If you are writing a note to your friends, you will use a more casual or informal voice, as if you were talking to them in person.

A. **Read each piece of writing below, paying attention to the writer's voice. Write who you think the audience is and the purpose for writing.**

1. I think I am an excellent candidate for student council president. I am a great listener. I like talking to people and getting to know them and listening to their concerns. I can make things happen. I never give up, and am committed to fighting for what I believe in and what my fellow students believe in.

    Audience: _____

    Purpose: _____

2. My new phone is garbage! The most annoying thing about it is that the touch screen doesn't work. I have to keep pressing the same spot over and over again! Also, the camera is terrible. The pictures are all grainy. I would not recommend this phone!

    Audience: _____

    Purpose: _____

B. **Write a review of a movie you have seen recently. Imagine that your review will be published in the school newspaper. Choose a voice that is appropriate for your purpose and audience.**

_____
_____
_____
_____
_____
_____

LESSON 70
STATE YOUR PURPOSE: TOPIC AND THESIS

When you write an expository essay, you need both a **topic** and a **thesis**. A topic is what the essay is about. A thesis is what you say about the topic, or the point you plan to make in your essay.

Your thesis should not simply be a fact; it should show a perspective on a topic and introduce a claim that will be supported in the rest of your essay.

For example: Topic: technology
Thesis: Spending too much time using electronic devices can have negative health effects.

It's important that your audience understands your purpose, so make sure your thesis states your topic clearly. Your thesis statement should answer the question "What is the purpose of my essay?"

A. Read the suggested thesis statements for each topic. Underline the thesis statement that you think does a better job of stating the purpose of the essay. Then, write why you think it is a better thesis.

1. Topic: natural resources

   a) Natural resources are minerals, forests, water, and land that are in nature.

   b) Preserving natural resources is one of the most important challenges facing the world today.

   _____
   _____
   _____

2. Topic: the War of 1812

   a) While the war of 1812 helped Canada develop a national identity, it also caused great losses for Indigenous people.

   b) The War of 1812 was a conflict between the United States and Great Britain.

   _____
   _____
   _____

B. Write a statement describing the difference between a topic and a thesis.

CLUSTER WITH A WEB: ORGANIZING IDEAS

One way to **organize ideas** to use for your writing is to cluster them in a topic/subtopic web. The topic in the main circle of your web tells the main idea. The subtopics in the smaller circles give more details about the main idea.

For example:

If you use a topic/subtopic web to help you organize an expository essay, the smaller circles should include the points that will support your thesis.

Organizing ideas using a topic/subtopic web helps you to plan what you will write about and to separate your writing into paragraphs or sections that focus on one idea.

A. **Fill in the topic/subtopic web below about how humans impact the environment. Write your main idea at the top of the web and think of three subtopics related to the main idea.**

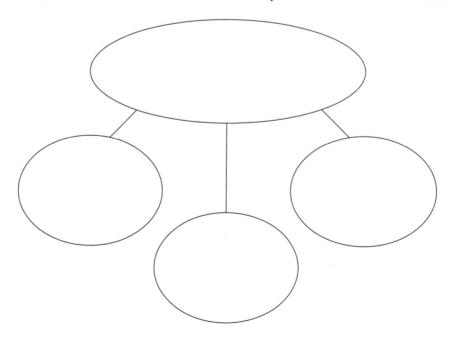

B. **Create a topic/subtopic web about a subject you want to write about in science or social studies. Organize your ideas into main idea and subtopics.**

USE A GRAPHIC ORGANIZER: ORGANIZING IDEAS

When you write, it is helpful to **organize your ideas**. There are many ways to organize ideas, each with a different purpose. The trick is to find the right way to organize ideas for your purpose.

A good way to organize ideas is to use a graphic organizer, such as a process chart. A process chart shows the different steps in a process, in the order in which they happen.

For example: Topic: How to Write a Research Essay

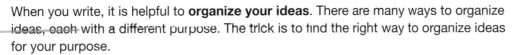

Another type of graphic organizer is a chronological order chart. This chart shows a sequence of events in the order in which they happened. A chronological order chart is useful when writing about a historical event or a news story.

For example: Topic: The Red River Resistance

A. Decide whether a process chart or a chronological order chart would be best for each writing topic below.

1. the plan for a group science experiment _____
2. a biography of Laura Secord _____
3. a news article about a robbery _____
4. instructions on what to do in case of a fire _____

B. Create a process chart about how you prepare for a test. Write each step in your test preparation routine and connect the steps with arrows.

C. Create a chronological order chart about your life so far. Write memorable or major events in the order in which they happened.

D. Use the information from your graphic organizer in Exercise C to write a paragraph about the memorable or major events in your life so far.

_____
_____
_____
_____
_____
_____
_____
_____
_____
_____
_____
_____
_____
_____
_____
_____

E. Choose a topic that you are interested in writing about. Decide what type of graphic organizer you would use to organize your ideas on the topic. Write why your chosen type of graphic organizer would work best for your purpose.

LESSON 73: USE DIALOGUE: STRONG OPENINGS

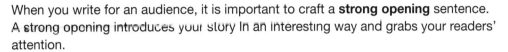

When you write for an audience, it is important to craft a **strong opening** sentence. A strong opening introduces your story in an interesting way and grabs your readers' attention.

One way to craft a strong opening is to use dialogue. Opening a narrative with dialogue introduces your readers to an event that is in progress. It is most effective if the dialogue introduces a conflict or a suspenseful situation.

For example: "Leave me alone!" I shouted, slamming the door behind me. This was the last straw. I was convinced I would never speak to my brother again.

"Watch out!" I called out to Jacinto. He looked up just as the model airplane came barrelling past his head.
"That was close!" he shouted.

Remember to punctuate your dialogue correctly. Put punctuation like exclamation marks inside quotation marks.

A. Write a strong opening with dialogue for each writing idea.

1. a disagreement between two friends

_____
_____
_____

2. trying to find a lost pet

_____
_____
_____

3. getting lost in the woods

_____
_____
_____

B. Find one of your favourite short stories. Does the first sentence include dialogue? If so, write whether you think it is an effective opening, and why. If not, rewrite the first line to include dialogue, and compare the two versions. Which version do you like better? Why?

LESSON 74
LEAD WITH A STATISTIC: STRONG OPENINGS

The opening for a piece of nonfiction writing is important. It should grab your readers' attention and make them want to continue reading.

One way to create a **strong opening** is to lead with a statistic. The statistic should capture your readers' interest. The sentences that follow should explain the statistic and identify the topic for readers.

For example: Did you know that more than one billion messages are sent through Facebook every day? Social media use is continuously on the rise across the globe.

In this example, the writer uses a surprising statistic to introduce the topic of social media. The statistic can be phrased as a question, a statement, or even an exclamation.

A. Write a strong opening sentence for each of the following statistics. Follow each opening sentence with a sentence that identifies the topic.

1. 61 000 Canadian soldiers killed in World War I

   _____
   _____
   _____
   _____

2. One in four Canadians is obese.

   _____
   _____
   _____
   _____

3. carbon dioxide in the atmosphere up 25 percent since 1958

   _____
   _____
   _____
   _____

B. Start a list of interesting or surprising statistics that you hear or read about. Use your list as a reference for creating strong openings in your writing. Think about how you can write the information in the most powerful way possible.

USE YOUR SENSES: WRITING DETAILS

Writers use descriptive **details** about characters and settings to make a story come alive for readers.

Descriptive details are often adjectives; they are the words that describe a person, place, or thing. The more descriptive details included in your story, the better readers are able to visualize character and setting. One way to write descriptive details is to use your senses. This will help your readers to see, hear, feel, taste, or smell what you are writing about.

For example: The beach was hot. ✗
I could feel the burning sand beneath my feet and smell the salty ocean air. ✓

A. **Use your senses to rewrite each situation with more detail.**

1. Inez was nervous.

_____

_____

2. The concert was loud.

_____

_____

3. The forest was dark.

_____

_____

B. **Use your senses to write a detailed description for each suggested topic.**

1. scientist

_____

_____

_____

2. animal shelter

_____

_____

_____

C. **Write two paragraphs describing one of your favourite memories. Use your senses to write details about the people and places.**

LESSON 76
USE EXAMPLES: WRITING DETAILS

Nonfiction writing always has a main idea and supporting **details**. The supporting details are smaller ideas that support the main idea.

One way to write supporting details is to write examples that describe, explain, or prove the main idea.

For example: Main idea: Canada underwent great political changes in the first half of the nineteenth century.
Supporting details: For example, the War of 1812 led to alliances between Indigenous and British forces. It also established a new-found sense of national identity in Canada.

Supporting details help readers to understand and visualize the main idea.

A. Read each paragraph. Underline the examples that support the main idea.

1. Part of the reason for climate change is the amount of greenhouse gases released into the atmosphere. Greenhouse gases come from several different sources. For example, carbon dioxide is released into the atmosphere naturally from humans and plants. It is also released through unnatural sources, such as the burning of fossil fuels like coal, oil, and natural gas. Other sources of greenhouse gases include livestock, wetlands, and landfills, which release methane into the atmosphere.

2. Being a student can be stressful. There are several ways for students to cope with stress. One way is to talk to someone, like a parent or a teacher. Talking to someone can help you work together to come up with ways to relieve stress. Even just having someone listen can make you feel better. Another way to deal with stress is to take deep breaths. This will help you relax and focus.

B. Write examples that support each main idea.

1. Physical activity has many health benefits.

   _____
   _____
   _____

2. My school has lots of different extracurricular activities.

   _____
   _____
   _____

C. **Write a short report about your favourite book, sport, or type of music. Start with the main idea (for example, why it is your favourite), and then list examples that support the main idea.**

LESSON 77

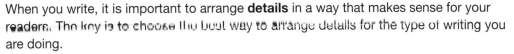
USE VARIETY: WRITING DETAILS

When you write, it is important to arrange **details** in a way that makes sense for your readers. The key is to choose the best way to arrange details for the type of writing you are doing.

Here are some different ways to arrange details in your writing. Each method of organizing ideas works well for different types of writing.

For example:
- Time: arranging details by time works well when you are describing an event.
- Location: arranging details by location works well for describing a setting.
- Importance: arranging details by importance works well for presenting facts in nonfiction writing.
- Compare and contrast: using a Venn diagram or a compare and contrast chart helps you to compare two things or ideas in nonfiction writing.

A. Read the following writing topics. Then, underline the method of arranging details you would use for each topic. Explain why you chose that method.

1. a description of the dining room at a glamorous restaurant

    time   location   importance   compare and contrast

    _____
    _____
    _____

2. a paragraph about what happened on your last day of school

    time   location   importance   compare and contrast

    _____
    _____
    _____

3. an article about how to avoid sunstroke and sunburns on a hot summer day

    time   location   importance   compare and contrast

    _____
    _____
    _____

B. **Choose one of the writing topics and underline your choice. Then, circle how you will arrange the details in your writing. Write some details about the topic and arrange them based on the method you chose.**

Topics:

a memorable day	my best friend's house	an event in Canadian history

comparing two types of renewable energy	my favourite sport	advice on what to do in an emergency situation

Method of Organization:

time	location	importance	compare and contrast

_____
_____
_____
_____
_____
_____

C. **Write a paragraph about the topic you chose in Exercise B.**

_____
_____
_____
_____
_____
_____
_____
_____
_____
_____
_____
_____
_____
_____

LESSON 78: FORMAT A SPEAKER'S WORDS: WRITING DIALOGUE

Writers use **dialogue** to tell what characters say, think, and feel. It is important to format dialogue correctly. There are several rules to follow when formatting dialogue.

Place quotation marks around the speaker's words and a speaker's tag before or after the speaker's words.

  For example: Glyn caught up with Shivani in the hall after class. "Shivani!" he called out. "Wait up!"

If the speaker tag comes before the speaker's words, place a comma after the speaker tag.

  For example: Shivani turned around and warned, "For the last time, I'm not singing in the talent show with you. I hate singing in front of people!"

If the speaker tag comes after the speaker's words, place a comma between the last spoken word and the end quotation marks.

  For example: "I don't know why you think I'd ask you to sing in the talent show," said Glyn.

If the speaker's words end with a question mark or exclamation mark, you don't need a comma.

  For example: "You've asked me every day this week! The answer is still the same: *no*!" Shivani replied.

A. Write the missing punctuation in each paragraph.

Peyton and Jonah were walking home from school. Did you hear about that earthquake? asked Peyton. I wish there was something I could do to help.

 Yeah, me too said Jonah. Maybe we can hold a fundraiser to help the relief efforts.

B. Finish this story. Include dialogue and format it correctly.

Last night, something strange happened. My brother Sam and I were watching a movie, when suddenly everything went dark. Then, our dog Kai started barking at something outside. "It's probably just a raccoon," said Sam. "I'm sure the power will be back on soon." He was trying to reassure me, but I could see that he was scared, too.

_____

_____

_____

_____

_____

C. Write a story about someone being granted one wish. Include dialogue and format it correctly.

D. In your own words, write notes about how to format dialogue correctly. Use your notes when you write dialogue in a story.

LESSON 79: MAKE LANGUAGE PRECISE: AVOIDING REDUNDANCIES

Unnecessary words or phrases can make your writing dull and repetitive. Using precise language will help make your writing clearer and more engaging for readers.

One way to make language precise is to **avoid redundancies**. Redundancies are words or phrases that are repetitive and add nothing to what has already been said.
   For example: The teacher gave us no *advance warning* about the math quiz.

In this example, the word *advance* is redundant because *warning* means to give advance notice.

A. **Read each sentence. Draw a line through each word that is redundant.**

   1. Later tonight, we will find out the end result of the election.
   2. Marnie and Jolene wore the exact same shirt to school today.
   3. Prince Edward Island and New Brunswick are joined together by the Confederation Bridge.
   4. Please proceed ahead to the front of the line.
   5. Juan wouldn't tell me the reason why he was late.
   6. We need to plan ahead for the school fundraiser next month.

B. **Rewrite this paragraph using more precise language. Remove any words in the underlined phrases that are unnecessary or repetitive. In one case, the verb following an underlined phrase changes tense when you remove the redundant word.**

My dad and I led the way <u>down to</u> the lake as my little brother <u>followed after</u> us. When we got there, we <u>looked out</u> across the frozen ice. I <u>knelt down</u> to <u>lace up</u> my skates, and then helped my brother with his. <u>Together, we</u> walked onto the ice and started gliding slowly. The <u>weather conditions</u> were perfect. In all of my <u>past experiences</u> on the lake, I had <u>never before</u> seen the ice so smooth.

_____
_____
_____
_____
_____

C. **Go on a search for redundancies. Look through magazines and newspapers to find at least three examples of redundancies. Start a list of redundancies that you can refer to when you are editing your writing.**

LESSON 80
SUM UP YOUR NARRATIVE: STRONG CONCLUSIONS

When writing a narrative, it is important to craft a **strong conclusion**. To be effective, your final sentence or paragraph should remind readers of the main topic or idea and let them know that the story has come to an end.

One way to write a strong conclusion is to compose a unique ending that will stick out in readers' minds. Ending with a strong emotion that draws from the events described in your story is a good way to make an impact on readers.

For example: At the end of the day, my cousins and I were tired but happy. ✗

As we drove home, I thought about all the amazing adventures we'd had over the summer. Then, a wave of sadness overtook me; this would be the last time I would see my cousins for a year. ✓

A. Read each description. Then, drawing from the emotions that the main character might be feeling, write a strong conclusion for each one.

1. Isaiah couldn't find his favourite hockey jersey, and he was convinced that his little sister had stolen it. After getting mad at his sister, Isaiah found his hockey jersey under his bed.

   _____
   _____
   _____

2. Yolanda spent the summer with her grandmother in Vancouver. She had never seen the ocean before, and they went fishing, whale watching, and kayaking. Yolanda didn't want to go home.

   _____
   _____
   _____

3. The day Lucy was supposed to give her speech at school, she realized she'd forgotten her notes at home. She decided she had to improvise. Although her speech ended up being different from the one she had written, she got a standing ovation.

   _____
   _____
   _____

B. Write three sentences about your idea of what makes a good ending to a narrative. Then, look through some stories and find an example of a strong conclusion. Write why you think that ending is effective.

LESSON 81
SUM UP YOUR REPORT: STRONG CONCLUSIONS

Just as it is important to begin a piece of nonfiction writing with a strong opening, it is also important to craft a strong conclusion.

One way to write a strong conclusion is to leave readers thinking about what they have just read. An effective conclusion will get readers to reflect on the topic or consider a perspective on the topic that they had not thought of before. Readers should take something away from the report that will stick out in their minds in the future.

To write a conclusion that leaves readers thinking, try ending with some advice, something you want readers to remember, or something you want readers to do.

For example: The next time you see someone litter or harm the environment, don't just stand there: say something. Spreading awareness of the problem and standing up for what you believe in plays a huge part in making a positive change in the world.

A. Read each topic and details below. Then, write a strong conclusion for each report.

1. Topic: the importance of natural resources

   Main points: Overuse of natural resources can lead to the destruction of forests and habitats, and can also contribute to climate change.

   Conclusion: _____

   _____

   _____

2. Topic: the impact of technology on the environment

   Main points: Transportation such as cars and planes cause air pollution; computers and other electronic devices can leak toxins into the environment if not recycled properly.

   Conclusion: _____

   _____

   _____

3. Topic: the importance of recycling

   Main points: Recycling used materials reduces the burden on our environment and helps preserve our natural resources.

   Conclusion: _____

   _____

   _____

B. Write a strong conclusion for this report.

Soccer is a very popular sport in Canada. However, all of the physical activity required to play soccer can lead to injuries.

    Soccer involves a lot of running and physical contact with other players. This contact can lead to bad bruises from hitting someone too hard or even a concussion if you hit your head. If you don't follow proper safety procedures and wear the suggested safety equipment, you are risking hurting yourself or others. Wearing proper safety equipment, such as cleats and shin guards, will help keep you safe.

    It is important to warm up properly before a game to avoid the risk of a sprain or a pulled muscle. If you hurt yourself during a game, it is important to address any injury you have, even if it feels small. Playing through an injury will make it a lot worse!

_____

_____

_____

C. Write a short report about a sport you either love or hate. Write a conclusion that will leave your readers with something to think about.

_____

_____

_____

_____

_____

_____

_____

_____

_____

_____

_____

_____

_____

_____

_____

LESSON 82: CATCH YOUR READERS' ATTENTION: EFFECTIVE TITLES

Crafting **effective titles** is an important part of writing. A title is the first thing about your text that readers will see, so it should catch their attention.

An effective title tells readers what your writing will be about. It should also be catchy and make them want to keep reading.

For example: All about the Underground Railroad ✗
The Underground Railroad: Path to Freedom ✓
My Trip to Montréal ✗
Bonjour, Montréal! ✓

It is important to format a title correctly. Capitalize the first and last word in your title, and every word in between except articles (*a*, *an*, and *the*), prepositions (such as *of*, *at*, *to*, *in*, and *about*), and conjunctions (such as *and*, *but*, and *or*).

A. Put a check mark beside the most effective title.

1. a) All about Structures
   b) Super Strong Structures
   c) Structures

2. a) Pond Wars! Bacteria in Ponds
   b) The Role of Bacteria in Ponds
   c) Ponds Have Bacteria

B. Rewrite the following titles to make them catchier. Format your titles correctly.

1. Green Vegetables Are Good for You

   _____

2. We Need to Turn Off Our Cellphones More

   _____

3. March Break

   _____

4. What I Love about Summer

   _____

5. Strength-Training Apps

   _____

C. Choose a book or short story with a title that you think could be better. Explain why you think the title is not effective. Suggest a title that you think would be more suitable.

LESSON 83: CHECK FOR ERRORS: REVISING AND EDITING

After you have written a first draft, it is time to **revise** and **edit** your work.

There are a number of errors to watch out for when you revise and edit your writing. It is important to make sure that you use a variety of sentence lengths to make your writing flow smoothly. You should also check for correct capitalization and punctuation.

This checklist can help you revise and edit your work:

- Did I use a variety of short, medium, and long sentences?
- Did I capitalize the beginning of each sentence and each proper noun?
- Did I use correct end punctuation?
- Did I use commas in a series, in dialogue, and in compound sentences?
- Did I use quotation marks to show dialogue?

Revising and editing your work is an important part of the writing process. To help you check if your writing is smooth and free of errors, try reading your writing aloud. You can often hear errors before you see them.

A. Correct the errors in capitalization and punctuation in each sentence. Cross out the incorrect letter or punctuation mark and write the correction above it, if necessary.

1. I thought the speech on sustainability, was really interesting.
2. ashton and mara's plane arrived in ottawa an hour late.
3. My sprained ankle got worse because I kept playing soccer after I injured it
4. I wish I had studied harder for the history test, said Farrah. I don't think I did very well.
5. I ran as fast as I could but I didn't catch up to my brother.
6. We learned about the diet living conditions and gender roles of people living in New France and British North America in the eighteenth century.
7. "Remember to study for the final geometry test" said Ms. Marble as we walked out of class.

B. Rewrite the sentences below, varying the lengths of the sentences. You may add, delete, and change words and punctuation.

Shane got in trouble because he was talking during the assembly but it wasn't his fault. He was just excited because he had received the good news that he had won two movie tickets in a contest.

_____
_____
_____

C. **Revise and edit the paragraph below, checking for a variety of sentence lengths and correct capitalization and punctuation.**

Last month at school we had our vaccinations. I was nervous, I don't like getting needles. The last time I got a needle I became lightheaded and had to lie down. I knew i had to get it done. I told myself to be brave. They called my name. I sat down. I told the Nurse I was nervous. She told me to relax. She started chatting with me. What careers are you interested in? she asked. I said "I don't know." Then she asked me about my plans for the summer. before I knew it, it was over! All I felt was a little poke. The nurse had distracted me so much that I hardly even felt it. Ever since then, I have wanted to become a nurse.

D. **Write a paragraph about what a house in the future might look like. After you write your paragraph, revise and edit it.**

CORRECT SENTENCES: REVISING AND EDITING

One of the last steps of the writing process is **revising** and **editing**. Editing your work is an important part of the writing process. Editing your writing gives you an opportunity to make sure it is free of errors.

Make sure you check sentence structure when you edit your work. It is important to find and fix any sentence fragments and run-on sentences. Sentence fragments are groups of words that do not express a complete thought. Run-on sentences are two sentences that are joined together without the correct punctuation or connecting word.

For example: The next house on the left. (fragment)
Madhu lives at the next house on the left. ✓
I finished the race it was hard. (run-on)
I finished the race. It was hard. ✓

A. **Correct the errors in the following paragraph. Check for errors in sentence structure (fragments and run-on sentences). Rewrite the revised paragraph.**

My grandmother's house was always full of old trinkets and antiques. Whenever I visited her, I liked to look through her old treasures. She would show them to me and tell me the stories behind them. Like stepping back in time. "This was a clock your grandfather bought after the war," she'd say, holding up an old wooden clock that was missing a hand. My favourite thing in her house was her collection of china dolls. They were kept in a locked glass display case, every now and then she would open it up and let me touch them. There was one that had long, curly hair. She said it looked like me.

B. Write several paragraphs about a childhood memory. When you are finished, check your writing for run-on sentences and sentence fragments. Correct any errors that you find.

SECTION REVIEW

A. **Write who the most likely audience is for each writing assignment.**

1. a research paper on the effects of greenhouse gases

    Audience: _____

2. a short story that features talking animals

    Audience: _____

B. **Decide whether a process chart or a chronological order graphic organizer would be best to use for each writing topic below. Write the type of graphic organizer on the line.**

1. assembly instructions for constructing a model airplane _____

2. a news article about a community event _____

3. a history essay about the War of 1812 _____

C. **Use dialogue to craft a strong opening for each story topic.**

1. getting stuck in an elevator with your favourite celebrity

    _____

2. finding out that your two best friends planned a surprise party for you

    _____

3. discovering that you have been transported back in time to the 1800s

    _____

D. **Use your senses to write descriptive details about the characters and settings below.**

1. a scientist on a submarine

    _____
    _____
    _____
    _____

2. chef at a busy restaurant

    _____
    _____
    _____
    _____

E. Write at least two examples that support the main idea below.

Main idea: Everyone can do his or her part to help the environment.

_____

_____

_____

_____

F. If you were completing each writing assignment below, how would you arrange the details? For each one, underline the best way to arrange the details.

1. an article describing life in Canada in the 1800s and life in Canada today

    time   location   importance   compare and contrast

2. a biography of Louis Riel

    time   location   importance   compare and contrast

3. a description of a haunted manor

    time   location   importance   compare and contrast

G. Add the missing punctuation to this dialogue.

We're almost there said Lewis as we trudged up what felt like an Everest-sized cliff. We were hauling all of our gear on our backs: tent, tarp, cooler, backpacks. I felt like my back was about to break into little pieces. Finally, we reached the top of the cliff.

Look over there, he said, pointing off in the distance. Do you see it?

Way over there? I thought you said we were close! Just then, we heard a clap of thunder.

I looked up. What had been a clear blue sky had turned dark and cloudy.

Better keep moving said Lewis, walking swiftly ahead of me.

H. Cross out the redundant word in each sentence.

1. Come into the store today and receive your free gift!
2. I have no future plans to become a mathematician.

I. Choose one of these writing topics. Then, write a strong conclusion for an article that will leave readers thinking.

the dangers of texting while driving   the results of overfishing   sports-related injuries

_____

_____

J. Underline the title in each set that is most likely to catch readers' attention.

1. a) Astronomy
   b) The Mysteries of the Sky
   c) All about Stars

2. a) Life below the Surface
   b) Ocean Life
   c) Sea Creatures

K. **Use the space below for a story outline. Then, on another piece of paper, write a short story about inventing a time machine. Include a strong opening that starts with dialogue and a conclusion that involves a strong emotion. Revise and edit your story.**

L. **Write one paragraph about an important Canadian. Include a title, a strong opening, supporting details, and a strong closing.**

DEVELOP RESEARCH SKILLS

Research is not just what you do to find an answer; it is an essential part of learning and an important part of the writing process.

The methods you use for your research will affect the quality of the information you get, whether you are looking up data to help you understand an issue, or searching for photos to help you illustrate a point. Good information makes good writing.

In this section, you will learn about the skills that make an effective researcher.

> "You have to do the research. If you don't know about something, then you ask the right people who do."
> — Spike Lee

LESSON 85: HAVE A CLEAR FOCUS: INQUIRY QUESTIONS

A great research project begins with an effective **inquiry question**. What makes an inquiry question effective?

First, an effective inquiry question is *meaningful* to you. You are interested in your topic, and you do not already know the answer to your question.

Next, an effective inquiry question is *open-ended*, meaning there is no simple answer. In fact, there may be several parts to the answer.

Often, effective inquiry questions are *debatable*, meaning that two people may have different opinions. Effective inquiry questions also require *research* to help you find your answer.

Finally, an effective inquiry question needs a *clear focus*. Questions that are too general can be difficult to manage.

   For example: What impact did the War of 1812 have on Canadian identity?

This is a good inquiry question because it has a clear focus, is open-ended, debatable, and requires research.

A. Read each of the following inquiry questions. On the line beside each question, write whether the inquiry question is *effective or ineffective*. On the lines below each question, explain why you made each choice.

   1. What is a mountain? _____

   _____
   _____

   2. How do the environmental challenges faced by settlers in 1750 compare with those faced by Canadians today? _____

   _____
   _____

B. Read the following inquiry question. Rewrite the question so that it meets the criteria of effective inquiry questions. Then, explain why your revised question is effective.

   When did slavery exist in Canada?

   _____
   _____

C. Think of a topic that you are interested in exploring. Write an effective inquiry question for your topic. Explain what makes your inquiry question effective.

140   Develop Research Skills

LESSON 86: FIND SYNONYMS: RESEARCHING WORDS

Researching words, like synonyms, can help you find the perfect words for your writing and help you avoid using the same word over and over.

Synonyms are words that mean the same thing, or almost the same thing, as another word. The best place to research synonyms is in a print or online thesaurus. For example, if you look up the word *smell*, you might find an entry like this:

smell (noun): aroma, odour, perfume

To decide which synonym to use, think about the sentence in which you plan to use the word.
   For example: The *smell* of baking cookies filled the room.

In this sentence, the word *aroma* would be a better replacement than *odour* or *perfume*.

A. Underline the two synonyms for each bolded word.

   1. **bravery**   courage   recklessness   weakness   fearlessness
   2. **impede**   allow   thwart   interfere   encourage
   3. **quarrel**   agreement   dispute   tranquility altercation

B. Write two synonyms for each word.

   1. safe _____ _____   4. irate _____ _____
   2. happy _____ _____   5. explore _____ _____
   3. motivate _____ _____   6. murmur _____ _____

C. For each sentence, circle the synonym that would be the best replacement for the underlined word.

   1. Max stomped his foot and said, "Hey! That's not fair!" (shouted / whispered / mumbled)
   2. My dad didn't think my jacket was right for today's weather. (exact / appropriate / true)
   3. The antique china in the museum is very delicate and could break. (crumble / smash / gap)
   4. I was starting to think that no one believed my story. (consider / feel / suspect)
   5. Mohammed liked the movie, but Ryu disliked it. (shunned / hated / cursed)

D. Look through some of your writing and create a list of words you use frequently. Use a thesaurus to find synonyms for these words. Then, add them to your list for future reference.

LESSON 87
CHOOSE RESOURCES: LIBRARY RESEARCH

When you do **library research**, there are many resources you can use to help you find information. Many of these resources are available online, as well as in print.

A *library database* gives you the names of authors and titles of books related to specific topic searches. An *encyclopedia* provides general information on a variety of topics. A *dictionary* provides meanings, spellings, and pronunciations of words. A *thesaurus* provides synonyms and antonyms. *Newspapers* give updates on current events. An *atlas* contains maps and other geographical information. An *almanac* contains statistics about information such as weather, population, moon phases, and tide tables. A *manual* tells you how to do something or how something works.

Choose resources that are most likely to help you find the information you are looking for. For example, if you are researching a flood that is currently happening, you might refer to a newspaper. If you want to find out about the land affected by the flood, you might look in an atlas for a map. An almanac could tell you about weather patterns in the flooded area.

A. Read each of the following scenarios. Circle the best resource for the information needed.

1. Aria wants to find a landform map. Where should she look?

   a) thesaurus
   b) almanac
   c) manual
   d) atlas

2. Otis is looking for general information about the Gobi Desert. Where should he look first?

   a) encyclopedia
   b) atlas
   c) almanac
   d) newspaper

B. Draw a line between the type of information in the left column and the best place to find it in the right column.

current election	manual
average annual temperature of Vancouver	newspaper
maintain an ATV engine	library database
pronunciation of a word	almanac
book recommended by a friend	dictionary

C. Select a topic that interests you. Make a list of the different types of resources you might use to research this topic. Find two of the resources. Which one contains the best information about your topic?

LESSON 88: CONDUCT ONLINE RESEARCH: KEYWORDS

The Internet gives you immediate access to lots of information. However, you need to use it efficiently or you will be overwhelmed with too much information. To help you find exactly what you need, use precise **keywords** when you do research. The more precise your keywords, the more effective your search results will be.

Start by setting a research purpose. What do you want to find out?
  For example: Why did the British expel the Acadians from Canada?

Use your topic to help you choose the most important keywords. Use nouns whenever possible. Place the most important words first. Use as many words as needed to narrow your search.
  For example: Acadians (weak); Acadians Canada (better); Acadians expelled Canada (best)

Avoid using full sentences or questions as they may lead to unnecessary information. Always be precise and specific!

A. Choose precise keywords for each research question. Write the keywords in order of importance.

1. Who is the president of the World Wildlife Fund (WWF) in Canada? When was he or she made president?

   _____
   _____

2. What discrimination did Black Loyalists face in New Brunswick in the late eighteenth century?

   _____
   _____

3. What were the terms of the Treaty of Utrecht?

   _____
   _____

4. What is the environmental impact of overfishing in the Bay of Fundy?

   _____
   _____

B. Enter your keywords for one of the topics from Exercise A into an Internet search engine. Were you able to find what you were looking for within the first few results? Did you need to modify your search to yield better results? Write a brief summary of which keywords were the most effective.

LESSON 89
KNOW THE DIFFERENCE: PRIMARY AND SECONDARY SOURCES

To research a topic, you can use both primary and secondary sources. Each source gives you a different type of information.

A **primary source** is a first-hand account of an event or experience. An eyewitness's account of an event, time, or place is a primary source. An object, artifact, or artwork created during the period of time being studied is also a primary source.

To search for primary sources related to your research topic, include the words *primary source* in your keyword search. To search for a specific type of primary source, enter the specific form, such as *speech*, *interview*, or *diary*.

For example: *Battle Plains Abraham primary source; Battle Plains Abraham letter*

A **secondary source** is a second-hand account of an event or experience. It is written or created by a person who did not experience the event, time, or place directly.

A. Decide whether the listed sources are primary or secondary sources, and then place them in the appropriate column of the table.

data from an experiment

diary

encyclopedia

interview

biography

autobiography

book about the Seven Years' War

artifacts from an archaeological site

history book

Primary Sources	Secondary Sources

B. Imagine you are researching a historical event, like the Battle of the Plains of Abraham. Explain the benefits of using a secondary source such as an encyclopedia and the benefits of using a primary source such as a letter from a soldier.

_____

_____

_____

C. Look through an online article or encyclopedia entry about a topic you are interested in. What primary sources has the article or encyclopedia entry referenced? How do you know?

LESSON 90
LOOK FOR CONSENSUS: EVALUATING WEBSITES

When you **evaluate a website**, look for consensus among sources. In other words, look for other sources that confirm the information you have found. To help you determine if information on a website is reliable, ask questions such as the following:

- Does the site have information that can be found or confirmed elsewhere?
- Does the information include sources that have been carefully cited?
- Is there a bibliography I can cross-check?
- Are there links to other sites that I know are credible?
- Does the site show obvious opinions or biases?

Look for at least three reliable sources that verify or confirm the information you have found.

A. **Read each statement below. Put a check mark (✓) beside each statement that is correct.**

a) Everyone who posts on the Internet is an expert, so you can trust all sources. ___

b) Websites that present a personal opinion are fine to use in research, as long as you agree with the author's opinion. ___

c) Reliable sites give evidence for their statements and provide citations for their claims. ___

B. **Read the following statements. Answer the questions that follow.**

1. Environment Canada has a website about managing and reducing waste. It matches information you have found in many other sources, and it has links to other information and sites that you recognize. Should you trust the information? Why or why not?

_____
_____
_____
_____

2. A website provides the author's opinion about proposed changes to public transit. The author provides many statistics to back up her claims, but does not reference or cite her sources of evidence. Should you use them? Why, or why not?

_____
_____
_____
_____

LESSON 91
BE CRITICAL: EVALUATING WEBSITES

Not all websites are created equal: some contain fact, others contain opinion, and some contain information that is incorrect. Be discerning! **Evaluate websites** to determine which ones are reliable. Ask yourself questions such as the following:

- Is the author an expert on the topic?
- Do multiple sources support the information?
- Is the purpose of the information to inform, instruct, persuade, or sell?
- Is the purpose of the website clear?
- Does the information contain obvious errors?
- Does the information come from an educational, business, or personal site?
- Is there a time stamp? How recently was the information published or updated?
- Is it clear whether the website is stating facts or an opinion? Are opinions disguised as fact?
- Does the information include sources that have been carefully cited?

A. Read each scenario and answer the question that follows.

1. While researching the War of 1812, you discover a site that has some brief facts and statistics. Most of the site is devoted to selling replica clothing. Should you trust the facts and statistics on this site? Explain.

   _____

   _____

2. While researching the environmental impact of industrial pollution on a river system in Canada, you find a blog from an environmental interest group. The site does not provide sources of evidence, and you cannot verify some of the claims they make. Should you trust the information? Explain.

   _____

   _____

3. A website about middle-school chemistry is written by a chemistry specialist. A link to the author's background is provided, and the information on the site can all be confirmed with other sources. The site has a bibliography, but not a time stamp. Should you trust the information? Explain.

   _____

   _____

   _____

B. Write a paragraph explaining why it is important to evaluate websites for reliability.

LESSON 92: REMIX AND REWORK: PLAGIARISM AND COPYRIGHT

Plagiarism, the act of copying some or all of the work of someone else and presenting it as your own, is against the law. **Copyright** is the law that protects people from having their work stolen. Written text, photos, illustrations, music, videos, and all other creative works are protected by copyright law.

When you rework or remix creative materials, such as when you create an audio or video remix or change a photo using image-editing software, you must respect the creative works of others. Follow these guidelines to avoid plagiarism and to obey copyright law:
- Always credit the original creator.
- Use the material in a new or different way, or add meaning to it.
- Use the works for nonprofit purposes only.

These guidelines only apply to borrowing ideas for the purposes of education, news reporting, reviewing, and parody or satire. When you borrow ideas or creative material for other purposes, you must obtain the author's or the creator's written permission. Ask a teacher or librarian, or research *Canada's Copyright Act,* to find out specific details.

A. **Which of the following is NOT an example of reworking or remixing fairly? Circle your choice.**

1. Dmitriy makes a video remix using clips from 15 different songs and gives credit to the original creators at the end of the remix.

2. Elise takes a photo she found online, uses a software program to edit the image, and adds the image to her report. The image is very different from the original, so she does not credit the photographer.

3. Nikita rewrites the lyrics to a popular song to add humour to a presentation he is preparing. He credits the songwriter in the presentation.

B. **Read the following scenario and circle the correct answer:**

Yasmin creates a video explaining why she doesn't like horror movies. She uses clips from movies to support her views, and she credits the creators of each movie she includes. Yasmin can use the movie clips because

1. Horror movies are the worst genre of movie ever created.

2. Movies are not copyrighted.

3. Her video is a review.

C. **Cover the instruction box at the top of this page. In your own words, explain what plagiarism is.**

_____

_____

D. **Write a paragraph explaining why it is important to credit the creators of materials that you rework or remix.**

LESSON 93: CITATIONS FOR VIDEOS: CREDITING SOURCES

When you use information or clips from a video in your presentation, you must **credit** the video producer. This means you tell who created it and where it came from.

To **cite a video** in a bibliography, include the following features, with punctuation: creator's name (last, first) or poster's username. "Title of image or video." Media type. *Name of website*. Name of website's publisher, date of posting. Medium. Date accessed.
For example: Rivera, Daniel. "Ecotourism in Costa Rica." Online video clip. *YouTube*. YouTube, 1 Oct. 2014. Web. 24 Apr. 2015.

The author and username, as well as the date of posting, are usually located directly below the video on a website.

A. Create a citation for each video clip. Use punctuation and italics correctly.

1. Video title: Clear-Cutting Rainforests in Madagascar
   Poster's username: EarthMattersNow
   Name of website: YouTube
   Date of posting: 12 Nov. 2011
   Date accessed: 26 Jun. 2015
   Website's publisher: YouTube

   _____
   _____

2. Video title: Green Heroes: The Art of Change
   Creator's name: CineFocus Canada
   Name of website: Green Heroes TV
   Date of posting: 27 Dec. 2014
   Date accessed: 22 Sept. 2015
   Website's publisher: TVO

   _____
   _____
   _____

B. Explain why it is important to credit video creators.

   _____
   _____
   _____

C. Find an online video related to a topic you are studying. Prepare a citation for the video.

LESSON 94: TRACK PRINT AND ONLINE SOURCES: RESEARCH NOTES

When you prepare a research project or presentation, you must credit all of the sources that you use. As you take **research notes**, develop a system to keep track of sources so that you have enough information when it's time to create a bibliography. You can't use a great quote, an important fact, or an interesting video clip if you can't remember where you found the information.

Keep track of **print** sources by writing the author, title, and page number(s) beside notes as you write them.

Track **online** sources by creating a bookmark folder for your research topic in your web browser. Each time you visit a website about your topic and take notes, add it to your bookmark folder. Each source will be stored in the order you added it. As you write notes from each source, record the order of the bookmark link. For example, if your notes correspond to your second bookmark, write a 2 beside the notes.

If you use information from a particular source to write your report, use your notes or bookmark folder to create a proper citation for your project. For a written project, prepare a bibliography or works cited page. For a multimedia presentation, prepare a references slide as the last slide of your presentation.

A. **Read each scenario. Circle the correct answer.**

1. Rickesh prepares a written report. He uses his research notes, but he has not recorded where they came from. His teacher did not specifically ask for a bibliography. What should Rickesh do?

    a) Do nothing, since his teacher didn't ask for a bibliography.

    b) Write a note at the end of his report saying that he forgot to keep track of his sources.

    c) Go back and find where he got the information, and create a proper bibliography.

2. Molly prepares a multimedia presentation. She remembers to keep track of her sources as she researches. What should she include on the sources slide of her presentation?

    a) every source she used to help her create the presentation

    b) every source she looked at, including those she didn't use for the presentation

    c) only the sources she found most helpful for her presentation

B. **List three different types of sources that would require tracking and citing if used in a multimedia presentation.**

_____

_____

_____

C. **Choose an inquiry question from the list, or choose an inquiry question you are interested in researching. Find three resources to help you answer your question. Write a few notes from each source, and track each source using the methods described in the box at the top of page 149.**

How do human activities contribute to droughts?

How does our natural environment shape our Canadian identity?

What can you do to effectively reduce your environmental footprint?

___

D. **Think about the process you used to track notes in Exercise C. What worked well? What might you change? Write a paragraph explaining what you believe is the most methodical system to help you keep track of print and online sources as you research.**

LESSON 95: USE IDEAS AND WORDS: PARAPHRASING AND QUOTING

When you **paraphrase**, you use your own words to explain someone else's ideas. You share information in a new way instead of copying your research word for word from an information source.

Almost all of your research paper should be in your own words. Sometimes, though, you may want to **quote** someone. When you quote someone, you use their exact words. Quotations can be helpful when you want to
- show that an expert supports your point of view
- give a definition
- use specific words as evidence, for example, when writing about a novel

When you quote someone, you must put quotation marks around the words you are using, and include the last name of the person who said the words in parentheses after the quotation. If the information came from a print source, like a book, you must also include the page number.

> For example: Earthship buildings are environmentally friendly because when they are constructed, "builders try their best to reuse as many materials as possible" (Tate 29).

In your bibliography, include a full citation for every author whose ideas you have paraphrased and quoted.

A. Rewrite this information in your own words.

For the first ten years that he spent in Canada, General Isaac Brock wanted nothing more than to return to Great Britain.

_____

_____

_____

B. Read the following quote from page 52 of a book by Paul Gibbert. Write a sentence showing how you could quote the information in a report.

Quote: General Isaac Brock was a brilliant tactical planner who changed the course of Canadian history.

_____

_____

_____

C. **Read the following paragraph about droughts. Identify the main idea and select one sentence or part of a sentence that you may wish to quote in a report. After you have recorded a sentence to quote, write point-form notes, focusing on key facts and details.**

A drought is a period of several weeks, months, or even years without rain. Soil becomes dry and cracked, rivers and streams dry up, and many plants and animals die without water. The impact of drought can be devastating. In the 1930s, in a period now known as the Dust Bowl, a severe drought in parts of North America caused the nutrient-rich soil in fields to turn to dust. Strong winds created dust storms that blew most of the soil away. Farmers were forced to leave their farms and look for other work.

Main idea: _____

Possible quotation: _____

_____

Notes: _____

D. **Use your notes to write a paragraph about drought. Include the quotation, using proper formatting. Assume the author's name is Sam Wentworth and the information is taken from page 145.**

SECTION REVIEW

A. **Read each inquiry question. On the line beside each question, write "Effective" or "Ineffective." Explain why you made each choice.**

   1. What is a fad diet? _____

   _____

   _____

   2. Are some fad diets more dangerous than others? _____

   _____

   _____

B. **Match each word on the left with a synonym on the right. Draw a line between the matching words.**

   acrimonious   hesitate

   gambit   hostile

   dither   arrogant

   pompous   strategy

C. **In the following sentence, replace the crossed out word with a synonym:**

   The water was very ~~dirty~~ _____ from all the churned up mud.

D. **Match the information in the column on the left with the resource in the column on the right that would be most helpful in finding it. Draw a line between the information and the resource.**

   recent typhoon   library database

   better words for an essay   atlas

   repair a lawnmower   manual

   book recommended by a friend   newspaper

   topographical map of Canada   thesaurus

E. **Choose precise keywords for each research question. Write the keywords in order of importance.**

   1. Who is current minister of health for Alberta?

   _____

   2. How did the Canadian Pacific Railway affect the development of Canada?

   _____

Develop Research Skills

F. Read the various sources of information below. On the line beside each source, write either *P* for a primary source or *S* for a secondary source.

1. data from a survey _____
2. encyclopedia _____
3. autobiography _____
4. journal entry _____
5. fossil _____
6. book about history of Canada _____

G. Read each statement. Put a check mark (✓) beside each statement that is correct.

a) Firewalls on computers are excellent, so you can trust information you find on all websites. _____

b) If you find a website you believe is reliable, you should find consensus with at least two other sources before you use any information from it. _____

c) If you agree with the author of a website's point of view, you can trust the information the website provides. _____

d) Reliable sites give evidence for their statements and provide citations or a bibliography. _____

H. List three questions that you can ask as you view websites to help you evaluate their reliability.

1. _____
_____

2. _____
_____

3. _____
_____

I. Read the following scenario. On the line provided, write "True" or "False." Explain your choice.

Oscar creates a video remix for part of a school presentation. A friend offers to pay him for a copy of his remix. He creates a copy, being sure to include a page at the end of the remix that credits all the original producers. It is now fair for him to accept payment for the remix. _____

_____

_____

_____

J. Insert correct punctuation for the following citation for a YouTube video clip:

ScienceForTomorrow Producers, Consumers, and Decomposers Online video clip *YouTube* YouTube 22 Jan 2013 Web 3 Sept 2015

K. **Read the following paragraph. Identify the main idea, and then write notes focusing on key facts and details.**

Be thankful you weren't in the tiny town of Frank, Alberta, on the morning of April 29, 1903. At 4:10 a.m., a huge wedge from nearby Turtle Mountain broke away from the mountain, sending large boulders barrelling through Frank. This landslide, later named Frank Slide, lasted just 100 seconds, but was massive. The 1 km wide and 425 m high slab that fell left Frank in a state of disaster. The town, a large portion of the Canadian Pacific Railway, and a coal mine were covered in dust, rubble, and stone. The Frank Slide is one of the largest and deadliest landslides in Canadian history.

Main idea: _____

L. **Use the notes you wrote in Exercise K to write a paragraph about the Frank slide.**

M. Research a topic that interests you. Use at least two sources. Select one sentence or part of a sentence that you wish to quote. Write point-form notes focusing on key facts and details. Track your sources in the margin beside your quote and notes.

Topic: _____

_____

_____

_____

_____

_____

_____

_____

_____

_____

_____

N. Use your point-form notes from Exercise M to write a short paragraph about your topic in your own words. Include the quote, using proper formatting.

_____

_____

_____

_____

_____

_____

_____

_____

_____

_____

_____

_____

_____

_____

ANSWER KEY

Work with Vocabulary

Lesson 1—Use a Similar Word: Synonyms
A. 1. b 2. c 3. a
B. 1. base 2. classify 3. pressure
C. Answers will vary. Rewrite each sentence using a synonym for the underlined word. Each synonym should be appropriate for the context of the sentence.
D. Answers will vary.

Lesson 2—Use the Opposite Word: Antonyms
A. 1. foolish 2. unusual 3. investigate
B. Answers will vary. Provide an antonym for each underlined word. Antonyms provided should fit the context.
C. Answers will vary. Write one or two sentences for each pair of words, using both words in the sentence(s).
D. Answers will vary.

Lesson 3—Choose the Correct Spelling: Homophones
A. 1. No – wear 2. No – to 3. Yes 4. No – weak
B. 1. pour – pore 2. stationary – stationery
   3. there – their 4. heard – herd
C. Answers will vary. Write one or two sentences for each pair of homophones provided. Homophones should be spelled appropriately.
D. Answers will vary.

Lesson 4—Expand Your Vocabulary: Root Words
A. If necessary, consult a dictionary to help you find the root words. Most dictionaries provide a word history (etymology) at the end of each definition.
   1. *vis*; to see 2. *spec*; to look 3. *scrib*; to write
B. 1. False; the root word *gen* means "birth," which has nothing to do with eyesight.
   2. True; the root word *vid* means "to see."
   3. True; the root word *tele* means "far off."
C. Answers will vary.
D. Answers will vary.

Lesson 5—Understand Word Beginnings: Prefixes
A. 1. b 2. c 3. c

B. 1. interacting 3. interviewed
   2. antiwar 4. autograph

Lesson 6—Understand Word Endings: Suffixes
A. 1. de – -sion 3. insist 5. -graph
   2. e – insurance 4. persistence 6. de
B. division: the action of separating into parts
   adolescence: the state of being a teenager
   monograph: the written study of a single subject
C. Answers will vary. Some words with the suffix -*ance* are *acceptance*, *acquaintance*, *brilliance*, *clearance*, *disturbance*, *dominance*, *guidance*, *inheritance*, *performance*, *remembrance*.
D. Answers will vary.

Lesson 7—Combine Two Words: Contractions
A. 1. I'd 2. we 3. is 4. I 5. we'd 6. it's
B. 1. They had 3. She would
   2. They would 4. She had
C. 1. There's 2. It's

Lesson 8—Mind Your Meaning: Denotation and Connotation
A. 1. b 2. a
B. Answers will vary. *Sample answers:*
   1. Positive: Wow, your puppy is energetic.
      Negative: Wow, your puppy is hyper.
   2. Positive: That's a fascinating hat.
      Negative: That's an amusing hat.
C. Answers will vary.

Lesson 9—Use Strong Words: Nouns and Verbs
A. 1. Red River Colony 2. preserve 3. hostilities
B. 1. bilingual 2. contribute 3. informed
C. Answers will vary. Replace the underlined words with a more effective noun or verb.
D. Answers will vary.

Lesson 10—Use Colloquialisms: Informal and Formal Language
A. Answers will vary. *Sample answers*: What's up?: How are you doing?; I wasn't born yesterday: I am not completely naive or unaware; He seems to be feeling blue: He seems sad; Can I borrow a toonie, please?: Can I borrow two dollars, please?

B. 1. Answers will vary. *Sample answers*: I'd like to return my salad. The hair in it is making me feel ill.
   2. Could you turn your music down? It is annoying me.

C. Answers will vary. *Sample answers*: Yesterday, I saw <u>several</u> people gathered in the park. When I got closer, I saw that they <u>were watching</u> a <u>man</u> performing magic tricks. He was <u>competent</u>, but it <u>took little mental effort</u> to <u>understand</u> most of what he was doing. When the <u>man</u> passed around his hat and asked for <u>money</u> at the end of his act, I <u>left quickly</u>.

D. Answers will vary. Write two versions of a short conversation between two people arranging to meet each other later. The first version should be written informally, using several colloquialisms. The second should be written formally, without colloquialisms. Spelling, grammar, and punctuation should be correct.

Lesson 11—Use Variety: Figurative Language

A. 1. idiom   4. onomatopoeia
   2. personification   5. metaphor
   3. alliteration   6. simile

B. simile, personification, personification, alliteration and personification

C. Answers will vary.

D. Answers will vary.

Section Review

A.

Word	complete	accept	expand	awareness
Synonym	whole	welcome	grow	knowledge
Antonym	partial	refuse	shrink	ignorance

This exercise is a review for Lessons 1 and 2.

B. 1. seller, cellar   4. it's, its
   2. pail, pale   5. their, there
   3. your, you're

This exercise is a review for Lesson 3.

C. Answers will vary. Note that each word involves doing something from a distance.

This exercise is a review for Lesson 4.

D. Answers will vary. Possible responses include telecommunications, telepathy, telephone, televise.

This exercise is a review for Lesson 4.

E. That is the house we would have bought—let us stop and see if it has changed at all.

This exercise is a review for Lesson 7.

F. 1. opponent   2. guests   3. information

This exercise is a review for Lesson 8.

G. 1. Answers will vary.   3. realized
   2. Answers will vary.   4. neighbours

This exercise is a review for Lesson 9.

H. to drive someone up the wall–to irritate or frustrate someone; to put your money where your mouth is–to act in a way that supports your opinion; to pull someone's leg–to joke with someone; to be tickled pink–to be very happy

This exercise is a review for Lesson 10.

I. metaphor, personification, metaphor, idiom, idiom

This exercise is a review for Lesson 11.

J. Answers will vary.

This exercise is a review for Lesson 11.

K. Answers will vary.

This exercise is a review for Lessons 9 and 11.

Build Sentences

Lesson 12—Use Variety: Types of Sentences

A. 1. . – declarative   6. ? – interrogative
   2. . – imperative   7. . – declarative
   3. ! – exclamatory   8. ! – exclamatory
   4. . – imperative   9. ? – interrogative
   5. ? – interrogative   10. ! – exclamatory
                              or . – declarative

B. Last weekend, my dad took me and my friends camping. It was the best weekend ever! I had never been camping before, and I was worried that it would be boring. But it was so much fun! The first night, Dad scared us all. We were sitting around the campfire, and everything was quiet. Then, he said, "Did anyone hear that?" We shook our heads. "I swear, I just heard something," he said. "Look over there. Do you see anything?" Then, as we looked around, he snuck up on us and yelled, "Watch out!" We all screamed. He scared us half to death!

C. Answers will vary.

D. Answers will vary.

Lesson 13—Use Variety: Sentence Length

A. Answers will vary. Rewrite the paragraph using a variety of sentence lengths.

B. Answers will vary. Write a paragraph about a time when you worked very hard for something. Use a variety of sentence lengths in your paragraph.

Lesson 14—Combine Sentences: Compound Sentences

A. The answers in parentheses are correct, but may not convey the same meaning as the first conjunction.
   1. so (but)   2. but (and)   3. or (but)

B. 1. The tornado tore through town, and (so) it damaged a lot of houses.
   2. Fareed practised a lot for the tryouts, but (yet) he did not make the soccer team.

C. My mom and I were going to the art gallery, and I had been looking forward to it for weeks. She mostly does landscapes, but sometimes she does portraits too. We had bought our tickets in advance, so we could skip to the front of the line. My mom bought me a print of my favourite painting, and I put it up on my bedroom wall.

D. 1. Imani invited seven people to her party, but only five people could come.
   2. Aslan and Gita saw an owl perched on a tree by the side of the road, so they stopped the car to get a closer look.
   3. It might rain during the parade this weekend, or it might be sunny.
   4. Carla and I stayed up late working on our science project, and (*or so*) we finished it in time for the science fair.
   5. I woke up feeling sick this morning, so I stayed home from school today.

E. Answers will vary.

Lesson 15—Expand Sentences: Adding Details

A. 1. (quickly) angry   3. (carefully) memorable
   2. (gracefully) frightened

B. Answers will vary. *Sample answers*:
   1. The little girl noticed a broken kite in the tall maple tree.
   2. The ancient lady with the fake glasses was running quickly toward the little girl.
   3. The genuine politician argued her points with zeal and emotion.

C. Answers will vary.
D. Answers will vary.
E. Answers will vary.

Lesson 16—Edit Sentences: Run-On Sentences

A. 1. ✓   2. ✗   3. ✓   4. ✗

B. 1. Karina wasn't wearing safety goggles, so the teacher asked her to put on a pair.
   2. The speakers talked about cyberbullying. It is an ongoing problem.
   3. I want to become a doctor like my mom, but it takes a lot of hard work.

C. Answers will vary.

Lesson 17—Know Complete Subjects and Predicates

A. 1. My older sister (won first prize in the speech-writing competition).
   2. Meghan and her friends (thought of several ways to reduce their carbon footprint).
   3. The neighbour's dog (was digging in our yard again this weekend).
   4. The hydroelectric dam (disrupted the freshwater ecosystem around it).
   5. Kyle's parents (were proud of him for passing his geometry test).
   6. The bold raccoon (broke through the screen and started rummaging through the cupboards).
   7. The ideal summer vacation (starts with a jump in the lake).
   8. All runners (need to stretch properly before and after going for a run).
   9. My dog and I (play fetch in the park every day after school).
   10. Canadians (celebrate the birth of their nation every year on July 1).

B. Answers will vary. Write a complete predicate to go with each complete subject.

C. Answers will vary.

Lesson 18—Identify Who or What: Simple Subjects

A. 1. technology   5. book
   2. player   6. athlete
   3. earthquake   7. problem
   4. writer   8. bowl

B. Answers will vary.

Lesson 19—Identify the Action: Simple Predicates

A. 1. went   5. had known
   2. is receiving   6. were barking
   3. started working 7. was getting
   4. raised   8. had been digging

B. Answers will vary.
C. Answers will vary.

Lesson 20—Identify Sentence Parts: Direct and Indirect Objects

A. 1. Cesar sent (his grandmother) a card.
   2. My mom made (me) a costume.
   3. The teacher showed (us) a documentary about the War of 1812.
   4. A police officer wrote (my dad) a ticket for speeding.
   5. Caitlin drew (her sister) a picture of a dinosaur.
   6. The fierce dog showed (me) his teeth.
   7. I offered (my mom) the spaghetti.

B. Answers will vary.
C. Answers will vary.

Lesson 21—Recognize Independent and Subordinate Clauses

A. 1. S   2. I   3. S   4. I   5. I   6. S

B. Answers will vary.
C. Answers will vary.
D. Answers will vary.

Lesson 22—Combine Sentences: Complex Sentences

A. 1. The community needs to be consulted (before)
   2. (Because) they sometimes experience flooding.
   3. We look at old photo albums (whenever)
   4. We like to practise in the gym (unless)
   5. (Since) people were advised to stay off the roads.
   6. Owen likes to go for a walk in the morning, (if)
   7. (Although) they discovered what caused it. You may also correctly identify the second part of the sentence as a subordinate clause: *it wasn't until the following century that*.

B. Answers will vary. Place the subordinate clause before or after the independent clause.
C. Answers will vary.
D. Answers will vary.

Lesson 23—Recognize Clauses: Adjective Clauses

A. 1. when I saw a caribou
   2. why I was grounded
   3. whose purse was stolen
   4. where I grew up
   5. that is more than sixty years old

B. Answers will vary. *Sample answers:*
   1. Students *who study hard* do well in school.
   2. The fruit *that was left for weeks in the hot kitchen* had gone bad.
   3. The house *that burned down* is for sale.

C. Answers will vary.
D. Answers will vary.

Lesson 24—Recognize Clauses: Adverb Clauses

A. 1. (because) I forgot mine at school.
   2. (until) he plays at the concert on Friday.
   3. (After) I watched the documentary,
   4. (while) she waited for her mother to pick her up.
   5. (Since) it was getting late,
   6. (unless) I have too much homework.

B. Answers will vary.
C. Answers will vary.

Lesson 25—Edit Sentences: Sentence Fragments

A. 1. ✗   2. ✗   4. ✗   6. ✗

B. Answers will vary.
C. Answers will vary.

Lesson 26—Edit Sentences: Comma Splices

A. Esther Brandeau was a brave woman, she was the first Jewish person to set foot on Canadian soil.

Esther sailed to Québec in 1738, she said her name was Jacques La Fargue. Her identity was discovered, she refused to convert to Catholicism.

Answers will vary. *Sample answers:*

Esther Brandeau was a brave woman; she was the first Jewish person to set foot on Canadian soil. Esther sailed to Québec in 1738. She said her name was Jacques La Fargue. Her identity was discovered, but she refused to convert to Catholicism.

B. Answers will vary. *Sample answers:*
It was a major earthquake. Many families had to be relocated.

It was a major earthquake and many families had to be relocated.

It was a major earthquake; many families had to be relocated.

C. Answers will vary.

Section Review

A. 1. statement   3. exclamation
   2. command   4. question

This exercise is a review for Lesson 12.

B. Answers will vary.

This exercise is a review for Lessons 13 and 14.

C. Answers will vary. Correct the run-on sentences by either adding a comma and a conjunction or by separating them into two sentences.

This exercise is a review for Lesson 16.

D. 1. A large group of tourists ⟨gathered around to take⟩ ⟨pictures of the war memorial⟩.
   2. My dad ⟨took⟩ me out on the sailboat for the ⟨first time this summer⟩.
   3. The bus full of travellers ⟨stopped suddenly to avoid hitting the turtle in the middle of the road⟩.
   4. My dog ⟨never seems to realize that his tail is actually attached to the rest of his body⟩.

This exercise is a review for Lesson 17.

E. 1. woman ⟨complains⟩
   2. teacher ⟨told⟩
   3. class ⟨started⟩

This exercise is a review for Lessons 18 and 19.

F. 1. ⟨me⟩ some advice
   2. ⟨her grandmother⟩ a card
   3. ⟨me⟩ a note

This exercise is a review for Lesson 20.

G. 1. S   2. I   3. I   4. S

This exercise is a review for Lesson 21.

H. 1. we started singing at the top of our lungs
   2. Laura took notes
   3. we'll make it to the final round
   4. I closed my eyes

This exercise is a review for Lesson 22.

I. 1. who found my phone
   2. where my grandmother grew up
   3. when I sprained my ankle
   4. that I planted

This exercise is a review for Lesson 23.

J. Answers will vary.

This exercise is a review for Lessons 25 and 26.

K. Answers will vary.

This exercise is a review for Lesson 14.

L. Answers will vary.

This exercise is a review for Lesson 22.

Know Capitalization and Punctuation

Lesson 27—Use Capitals:
A Variety of Capitalization

A. 1. cape breton island
   2. the niagara treaty of 1764
   3. gen. james wolfe
   4. the north star
   5. national acadian day

B. 1. the québec act of 1774 guaranteed religious freedom for the roman catholics in that province.
   2. the city of brantford is named after joseph brant, who was a political leader of the mohawk nation.

C. 1. Africville was an African-Canadian community near Halifax.
   2. The Africville Genealogy Society published a book called The Spirit of Africville.
   3. Another book, called Last Days in Africville, was nominated for the Book of the Year for Children Award.

D. Answers will vary.

Lesson 28—Use a Dictionary:
Abbreviations

A. 1. a   2. d   3. b

B. The death of an animal means life for fungi, because as fungi decompose the organic material from the animal, they also consume nutrients from it.

C. Answers will vary.

Lesson 29—Identify Short Forms:
Abbreviations

A. 1. I am actually laughing out loud in real life right now!
   2. My thoughts exactly! That's why you're my best friend forever.
   3. To be honest, I don't care what we do tonight.

B. Answers will vary. Write one sentence containing several abbreviations, and then rewrite the sentence, avoiding abbreviations and using proper spelling, grammar, and punctuation. Identify the audience for each sentence.

C. Answers will vary.

Lesson 30—Use Variety: Commas

A. 1. "Let's review conduction, convection, and radiation," Mr. Pinder said.

2. Camilla asked, "Mr. Pinder, when we talk about conduction, can you please give us a definition?"

3. "Yes, I will," he answered, "as soon as Chantal, Samson, and Inez stop talking!"

B. "Well, I'd like to start today's meeting by thanking everyone for coming," the volunteer coordinator said. She continued, "I know it's a beautiful day, and we would all love to be outside rather than here. Because the weather is so nice, I'll try to get through our agenda as quickly as possible. Is it OK if we skip our break, or will people need one in an hour? When we finish discussing this first item, we can decide."

C. Answers will vary.

Lesson 31—Punctuate Dialogue: Quotation Marks

A. Answers will vary. Rewrite the sentences, replacing the text in brackets with appropriate titles. Each title should have single quotation marks around it. Ensure that the title in the second sentence is followed by a comma, a closing single quotation mark, and a closing double quotation mark.

B. "'I knew I would never feel lost again' was my favourite line in your story, Pradeep," Myra said.

Pradeep replied, "Thanks! I originally wrote 'I knew I would never *be* lost again,' but I changed it."

Myra commented, "Your story reminded me of the song 'Lost Stars.'"

C. 1. "'Falling Down' is my least favourite song right now," said Arianna.

2. "If I hear the song 'Go Around' one more time, I think I'll scream," she added.

3. Marie-Louise agreed, "Yes, I'd much rather listen to 'Frantic.'"

D. Answers will vary.

Lesson 32—Show Possession: Apostrophes

A. 1. a   2. b   3. b

B. 1. aunt's   3. deck's   5. loon's
   2. boat's   4. masts'   6. albatross's

C. Answers will vary.

Lesson 33—Join Independent Clauses: Semicolons

A. 1. Many environmentalists oppose oil pipelines; they believe pipelines damage water sources and the land.

2. Some Indigenous communities do not want oil pipelines on their land; pipelines can disturb sacred places and contaminate fish.

B. 1. Humans use water for many purposes besides drinking and cleaning; we also use it to produce electricity and to water crops.

2. Canada's mines produce gold, silver, and diamonds; they are also known for nickel, iron, and coal.

C. Answers will vary.

Lesson 34—Separate Titles and Subtitles: Colons

A. 1. "Good Try!": How to Be a Supportive Teammate

2. Take Charge of Your Health: A Nutrition Guide for Teenagers

3. Help Is Just a Phone Call Away: Local Support Resources

B. Answers will vary. Write four subtitles preceded by colons. Subtitles should begin with capital letters and contain more information about the titles.

C. Answers will vary.

Lesson 35—Add Less Important Information: Parentheses

A. No; The second sentence is as important as the first and third sentences. In fact, it leads directly into the third sentence.

B. First, preheat your oven to 350° C. (Remember to empty it first!) Then, gather all of the ingredients. (Feel free to replace the walnuts with pecans or almonds.) Grease a muffin pan with butter and set it aside. In a large bowl, beat the butter, and then add the eggs and sugar. Beat until smooth. In a medium bowl, combine the flour, baking powder, and salt. (By now, your oven should be at the correct temperature.) Add the dry ingredients and the milk into the egg mixture. Stir until the dry ingredients are just incorporated. (If you stir the batter too much, your muffins will not be light and fluffy.) Add the nuts into the butter. Spoon the batter into the muffin pan, and then bake for 25 minutes.

C. Answers will vary.

Lesson 36—Guide Readers: A Variety of Punctuation

A. 1. :   2. ;   3. ;   4. :

B. "'Big, Big Snow' is Essie and Yusuf's favourite song from the festival," said their mother.

C. "We saw a huge drum and dancers in regalia at the festival!" said Essie excitedly.

"We also listened to a storyteller," added Yusuf. "She told a story called 'The Last Nakoda.'"

The siblings had thoroughly enjoyed the festival dancers; they had loved the music, too. (Yusuf's favourite new food was breakfast bannock.) The brother and sister's vacation was off to a great start; the rest of their planned activities also sounded wonderful.

"Tomorrow we're going hiking, right?" asked Essie.

Yusuf answered, "That's right. Tomorrow is the hike. Wednesday is fishing day!"

D. Answers will vary.

Section Review

A. 1. On April 17, 1982, the Constitution Act came into effect.

   2. Queen Elizabeth II signed the document in the city of Ottawa.

   This exercise is a review for Lesson 27.

B. 1. No 2. No 3. Yes

   This exercise is a review for Lesson 29.

C. "Well, being a good actor involves a good memory, the willingness to take risks, and a love of good stories," Aya told the reporter. "You can attend theatre school, but sometimes experience is more valuable than knowledge."

   This exercise is a review for Lesson 30.

D. 1. "I think we should call our skit 'A Day in the Life of a Mountain Climber,'" Prem said.

   2. "'Tackling Mount Everest' has a nice ring to it, too," suggested Yazmeen.

   3. "How about calling it 'Climbing to Success' or something?" Oskar added.

   4. "Looks like we'll need a vote!" Prem said.

   This exercise is a review for Lesson 31.

E. 1. walkers and bikers'   3. Andreas's and Gert's
   2. dogs' and people's

   This exercise is a review for Lesson 32.

F. 1. Olivia has taken several good photographs lately; she especially likes the close-ups of flowers.

   2. She is thinking about entering her photos in a contest; the winner receives a new camera!

   3. Olivia asks her aunt which photos she should enter; she respects her aunt's opinion on photography.

   This exercise is a review for Lesson 33.

G. 1. Thunderstorms: Nature's Convection Ovens
   2. Gassing Up: Our Contribution to the Greenhouse Effect
   3. Hot, Hot, Hot!: Three Ways to Produce Heat

   This exercise is a review for Lesson 34.

H. For our experiment, we wanted to find the best way to keep certain foods fresh. (Canadians waste a lot of food every year because it goes bad.) We chose to use strawberries and various kinds of containers. (We used four kinds of containers, but you can use more or fewer.) We put the same number of strawberries in each kind of container and put all the containers in the fridge. We checked each container every day to see how fresh the strawberries were. (You can use another type of food, as long as it's one that doesn't stay fresh for very long.) We recorded the results in a log, and then turned the results into a graph. Our conclusion is that a lettuce keeper is the best container for keeping strawberries fresh. (Another conclusion we came to is that handling rotten strawberries is disgusting!)

   This exercise is a review for Lesson 35.

I. 1. 'I've decided that I want to be a poet.' announced Oban.

   "I've decided that I want to be a poet," announced Oban.

   2. He added, 'I read an article called "How to Become a Poet, 10 Simple Steps": it sounds pretty easy.'

   He added, "I read an article called 'How to Become a Poet: 10 Simple Steps'; it sounds pretty easy."

   3. I asked, "Have you read the poem (I Lost My Talk) by Rita Joe?"

   I asked, "Have you read the poem 'I Lost My Talk' by Rita Joe?"

   4. Oban answered, "No: I haven't. I'll look for : in the library tomorrow."

   Oban answered, "No, I haven't. I'll look for it in the library tomorrow."

   This exercise is a review for Lesson 36.

J. Answers will vary.

   This exercise is a review for Lessons 27, 30, and 31.

K. Answers will vary.

   This exercise is a review for Lesson 33.

Grasp Grammar and Usage

Lesson 37—Name the Person, Place, Thing, or Idea: Nouns

A. cruise, reflection, laptop, sediment, guilt, lifestyle, curiosity

B. 1. collective 3. proper   5. collective 7. proper
   2. common   4. common   6. common   8. collective

C. Answers will vary. Proper nouns should be capitalized.

D. 1. common; idea
    2. proper; place
    3. collective; person
    4. common; thing
    5. proper; person

E. Answers will vary.

F. Answers will vary.

Lesson 38—Show Ownership: Singular Possessive Nouns

A. 1. party's theme
    2. bus's seats
    3. Earth's resources
    4. Marcus's pen
    5. Alberta's climate
    6. library's laptops

B. 1. leaf's
    2. plants
    3. valley's
    4. inbox's
    5. instruments
    6. butterfly's
    7. vegetables

C. Answers will vary. The possessive form of *Canada* is *Canada's*.

D. Answers will vary.

Lesson 39—Show Group Ownership: Plural Possessive Nouns

A. 1. groups'
    2. Manitobans'
    3. knives'
    4. turkeys'
    5. sandwiches'
    6. provinces'

B. 1. P   2. S   3. P   4. P   5. S   6. P

C. 1. dogs' barks   2. tornados' paths

D. Answers will vary.

Lesson 40—Use Irregular Plural Possessive Nouns

A. 1. deer's
    2. Moose's
    3. Men's
    4. sheep's
    5. children's
    6. Bison's
    7. media's

B. 1. children's
    2. women's
    3. geese's
    4. oxen's
    5. people's
    6. men's

C. Answers will vary.

Lesson 41—Use Concrete and Abstract Nouns

A. 1. A   2. C   3. A   4. C   5. C   6. A

B. 1. kindness
    2. confidence
    3. experiences
    4. summer
    5. respect
    6. wisdom

C. Answers will vary.

D. Answers will vary.

Lesson 42—Identify Action, Auxiliary, and Linking Verbs

A. 1. researched
    2. arrived
    3. participate
    4. found
    5. emigrated

B. 1. auxiliary
    2. linking
    3. action
    4. auxiliary
    5. linking
    6. action
    7. auxiliary
    8. linking

C. 1. Answers will vary.
    2. were
    3. Answers will vary.
    4. Answers will vary, but examples are *felt*, *appeared*, *were*.
    5. had, have
    6. Answers will vary.

D. Answers will vary. Write a short story that includes examples of action, linking, and auxiliary verbs.

Lesson 43—Provide More Information: Verb Phrases

A. 1. was stretching
    2. will gather
    3. did learn
    4. had completed
    5. was entering
    6. am experimenting
    7. have worked
    8. do enjoy
    9. should be practising

B. Answers will vary. Verb phrases should fit logically with the rest of the sentence.

Lesson 44—Show When an Action Happens: Verb Tenses

A. 1. simple past
    2. simple present
    3. present progressive
    4. simple future
    5. past progressive

B. 1. grow
    2. discovered
    3. will learn
    4. were walking
    5. are helping
    6. realize

C. Answers will vary. For #1, the tense is future, so your sentences should be in the past, present, or present progressive. For #2, the tense is present progressive, so your sentences should be in the past, present, or future.

D. Our class visited a freshwater marsh not far from our school to observe the wildlife. We saw a variety of birds, such as herons and ducks. We tried to spot insects such as dragonflies and butterflies. Some plants we discovered were reeds and cattails. As for mammals, a muskrat or two made an appearance.

E. Answers will vary.

Lesson 45—Make the Past Tense: Irregular Verbs

A. 1. shook   3. hurt   5. felt   7. cut
    2. rang   4. rode   6. wore   8. drank

B. 1. shut   2. thought   3. slept   4. drove   5. let

C. Answers will vary. The past-tense verbs should be *fed*, *grew*, or *chose*.

Lesson 46—Use Present Perfect and Past Perfect Tenses

A. 1. had completed; PP
2. had landed; PP
3. have lived; PrP
4. had seen; PP
5. have constructed; PrP
6. has eaten; PrP

B. 1. had flipped
2. had completed
3. has experienced
4. had seen
5. have grown

Lesson 47—Match the Numbers: Subject-Verb Agreement

A. 1. S; Amelie
2. S; team
3. P; boaters
4. P; stories
5. P; Min-jun and Jakob
6. S; race

B. 1. are attending   2. is   3. is representing   4. were

C. Answers will vary. #1 requires a singular verb; #2 requires a plural verb; #3 requires a plural verb.

D. The corrected version is as follows:

My friends Arun and Ivy and I <u>are wondering</u> what we could do today. Arun suggests that we play his drums and electric guitar, but apparently he doesn't understand the suffering that our "music" usually <u>causes</u> his family. A ride on the bike trails near our apartment building <u>is</u> the best suggestion so far. The morning is sunny and clear, perfect for riding. After our parents <u>give</u> us permission, we gather our gear and pack lunches and water, and off we <u>go</u>. Through wooded areas and open meadows our strong legs <u>take</u> us. Arun <u>grumbles</u> only twice about our choice to be outside.

Lesson 48—Match the Subject: Linking Verbs

A. 1. place; is
2. advertisements; were
3. rink; has been
4. flags; are
5. trail; has been
6. Hurricanes and typhoons; were

B. Answers will vary, but the verbs should be plural or singular as follows:

1. plural   3. plural   5. singular
2. singular   4. plural   6. singular

Lesson 49—Replace Subject Nouns: Subject Pronouns

A. 1. I   3. You   5. we   7. He and I
2. It   4. You and she   6. they   8. I

B. 1. They   2. It   3. They   4. You   5. We

Lesson 50—Replace Object Nouns: Object Pronouns

A. 1. <u>those straws</u>; them   3. <u>Hannah</u>; her
2. <u>this song</u>; it   4. <u>Teo and me</u>; you

B. 1. them   2. it   3. her   4. him   5. me, us   6. you

C. Answers will vary.

Lesson 51—Show Ownership: Possessive Pronouns

A. 1. mine   3. his   5. ours   7. theirs
2. yours   4. hers   6. yours

B. 1. hers   3. ours   5. yours
2. yours   4. theirs   6. mine

C. 1. mine (or his, hers, theirs)   3. hers
2. ours (or theirs)

Lesson 52—Use Indefinite Pronouns

A. 1. something; was   4. Everything; is
2. other; seems   5. much; is
3. Several; have   6. many; fall

B. 1. others   3. somebody   5. each
2. both   4. Much

C. Answers will vary.

Lesson 53—Use Reflexive Pronouns

A. 1. myself   4. yourselves   7. itself
2. themselves   5. yourself   8. herself
3. ourselves   6. himself

B. 1. ourselves   3. yourselves   5. yourself
2. myself   4. themselves   6. itself

C. Answers will vary. Each of the sentences should use a different reflexive pronoun.

D. Answers will vary. You should include at least three different reflexive pronouns that correctly agree with their subjects.

Lesson 54—Make Pronouns and Antecedents Agree

A. 1. it   3. he   5. them   7. his
2. they   4. we   6. he

B. 1. P – her; A – Madison
2. P – it; A – the candle
3. P – they; A – Erik and Pavani
4. P – we; A – Julia and I
5. P – his; A – Jon
6. P – it; A – video
7. P – it; A – The phone

C. Answers will vary.

Lesson 55—A Variety of Pronouns and Antecedents

A. 1. she   3. her   5. us   7. it
2. themselves   4. them   6. he

B. 1. Mary Pratt is a Canadian painter, and I intend to do some research about her.
2. We have more items to add to our work safety display, but we don't really need them.
3. Mikayla and I could not stop ourselves from laughing at the cat videos.

C. Answers will vary.

Lesson 56—Write Descriptive Words: Adjectives

A. 1. restless, busy, airport
2. mature, Pacific, many, its, original
3. this, kind, laughing, his, peculiar
4. terrified, long, his, that, second, horror
5. That, modern, Inuit
6. mighty, angry, young, her, its

B. Answers will vary. Write two adjectives for each underlined noun. Each adjective should make sense within the sentence and describe the noun accurately.

C. Answers will vary. Write three appropriate adjectives to replace each adjective provided.

D. Answers will vary. Include a variety of sensory, descriptive adjectives and adjectives that help give specific information, such as demonstrative adjectives and possessive adjectives.

E. Answers will vary.

Lesson 57—Make Comparisons: Adjectives

A. 1. more beautiful, most beautiful
2. more amazing, most amazing
3. healthier, healthiest
4. brighter, brightest
5. guiltier, guiltiest

B. 1. shinier   4. thinner   7. lazier
2. most disappointed   5. quickest
3. more responsible   6. hottest

C. Answers will vary.

Lesson 58—Describe Actions: Adverbs

A. 1. contributed; verb   5. tall; adjective
2. happy; adjective   6. confidently; adverb
3. solved; verb   7. high; adjective
4. brightly; adverb

B. 1. quite, regularly, always
2. frequently, extremely, fairly, quickly
3. very, safely
4. incredibly, too

C. Answers will vary.

Lesson 59—Make Comparisons: Adverbs

A. 1. longer   5. most
2. most confidently   6. more safely
3. least   7. furthest or farthest
4. more positively

B. Answers will vary. Write one sentence with a comparative adverb and one with a superlative adverb.

C. Answers will vary.

Lesson 60—Write Descriptively: Adjectives and Adverbs

A. 1. adverb   3. adverb   5. adjective
2. adjective   4. adverb   6. adjective

B. Answers will vary. Add at least one adverb and one adjective to each sentence and use them logically.

C. Answers will vary.

Lesson 61—Show Relationships: Prepositions

A. 1. For, to, in
2. On, to, with, in
3. In, near, with, for, at, after
4. across, from, to, to, unlike

B. 1. Following   3. Within   5. except
2. until   4. since   6. down

Lesson 62—Recognize Phrases: Prepositional Phrases

A. 1. (with) three standing ovations; (before) we left.
2. (Over) three months; (for) their charities.
3. (through) the park; (at) a restaurant; (for) falafels.
4. (In) the souvenir shop; (with) the maple leaf.
5. (In) the fall; (with) the harvest; (on) her parents' farm.
6. (At) the bowling alley; (in) a row.
7. (in) the green coat; (by) the door; (for) thirty minutes.

B. Answers will vary.

C. Answers will vary.

Lesson 63—Describe a Noun: Participle Phrases

A. 1. Clutching the railing; (we)
2. waiting patiently by the gate; (dog)
3. Encouraged by the applause; (singer)
4. fallen on the ground; (apples)
5. Frozen in alert stillness; (deer)
6. Reading about her ancestors; (Alana)
7. posted online; (interview)
8. studying the stars; (girl)
9. chosen by the librarian; (stories)

B. Answers will vary. If you place the participle phrase at the beginning of the sentence, you should follow it with a comma. If you place it after the noun and it is restrictive, there should be no comma. The phrase should agree with the noun it modifies.

C. Answers will vary.

Lesson 64—Be Clear: Misplaced Modifiers

A. 1. My sister's plastic toys are scattered on the playroom floor.
   2. They nervously walked [or *walked nervously*] onto the stage to present the play they wrote.
   3. The boy who was expertly wakeboarding waved to his friends on shore.
   4. The farmer's antique tractor was next on the auctioneer's list.

B. 1. I almost lost all the fundraising money.
   2. I lost almost all the fundraising money.

C. Answers will vary.

Lesson 65—Be Clear: Misplaced and Dangling Modifiers

A. 1. The zookeeper wearing an orange vest fed the lions.
   2. The lions lounging in the shade of a tree pounced on the food.

B. Answers will vary. You will need to add and/or change the order of words in the sentence.

Lesson 66—Use Joining Words: Conjunctions

A. 1. since; S 2. so; C 3. and; C 4. Unless; S

B. 1. while   3. when   5. so
   2. After   4. Unless 6. since

C. Answers will vary. You can use conjunctions from the list in the box at the top of the lesson or others of your choice. Each sentence ending should relate to the conjunction and to the first part of the sentence.

D. Answers will vary.

Lesson 67—Express Emotions: Interjections

A. 1. Yikes,   3. Well,   5. Bravo!   7. Ha!
   2. Oh no!   4. Yes,   6. Hmm,

Answers may vary in punctuation. In some sentences, both interjections could possibly work, but choose the most logical and effective one.

B. Answers will vary. Choose logical interjections for your sentences and use correct and appropriate punctuation.

Section Review

A. 1. collective, concrete   3. proper, concrete
   2. common, abstract   4. common, concrete

This exercise is a review for Lessons 37 and 41.

B. 1. auxiliary 2. action 3. linking 4. action

This exercise is a review for Lesson 42.

C. 1. past   3. present
   2. present progressive 4. future

This exercise is a review for Lesson 44.

D. 1. <u>Badminton</u>; is   4. <u>committee</u>; acts
   2. <u>plaques</u>; honour 5. <u>items</u>; are
   3. <u>She</u>; doesn't

This exercise is a review for Lesson 47.

E. 1. Petra and Jackson helped *themselves* to the fruit.
   2. We saw a play last night and really enjoyed *it*.

This exercise is a review for Lesson 54.

F. 1. three, great, park
   2. fastest, land, tall, dry, African
   3. tricky, soccer, more
   4. Note: Although the lesson specifies four adjectives, there are six. My, favourite, Canadian, our, fun, multicultural

This exercise is a review for Lesson 56.

G. 1. quickly; adverb   3. surprising; adjective
   2. waited; verb   4. wobbled; verb

This exercise is a review for Lesson 58.

H. 1. Until, to   3. Following, across, beside
   2. on, before   4. within, over, in

This exercise is a review for Lesson 61.

I. 1. The sculptor who had been *carefully* working dropped the sculpture on his foot.
   2. The family's *delicious* dinner was prepared by the children.
   3. She tripped over the child's *purple* shoe.

This exercise is a review for Lesson 64.

J. 1. <u>Guided by our expert</u>; (we)
   2. <u>standing by the door</u>; (woman)
   3. <u>confused by the instructions</u>; (Meg)

This exercise is a review for Lesson 65.

K. Answers will vary.

This exercise is a review for Lesson 66.

L. Answers will vary.

This exercise is a review for Lesson 60 and 67.

Craft and Compose

Lesson 68—Create a Life Map: Choosing a Topic

A. Answers will vary.

B. Answers will vary.

Lesson 69—Choose Your Voice: Purpose and Audience

A. 1. Audience: Classmates who will be voting for student council president. Purpose: To tell classmates why you would be a good student council president.

2. Audience: People interested in buying a new phone/online product review site or forum. Purpose: To tell the general public not to purchase the same phone as the one you own.

B. Answers will vary. Write a movie review using the appropriate voice for your audience.

Lesson 70—State Your Purpose: Topic and Thesis

A. 1. b   2. a

Answers will vary.

B. Answers will vary.

Lesson 71—Cluster with a Web: Organizing Ideas

A. Answers will vary. Create a topic/subtopic web about how humans adversely affect the environment. Fill in the main idea at the top of the web and the supporting details in the lower circles.

B. Answers will vary.

Lesson 72—Use a Graphic Organizer: Organizing Ideas

A. 1. process chart   3. chronological order
   2. chronological order   4. process chart

B. Answers will vary.

C. Answers will vary.

D. Answers will vary.

E. Answers will vary.

Lesson 73—Use Dialogue: Strong Openings

A. Answers will vary.

B. Answers will vary.

Lesson 74—Lead with a Statistic: Strong Openings

A. Answers will vary. Write two complete sentences: one opening sentence with the statistic and one that identifies the topic. The opening sentences should be relevant to the topic and interesting enough to grab readers' attention.

B. Answers will vary.

Lesson 75—Use Your Senses: Writing Details

A. Answers will vary. Use your senses to rewrite the descriptions with more details.

B. Answers will vary. Use your senses to write detailed descriptions of the character and setting provided.

C. Answers will vary. Write two or more paragraphs about a favourite memory. Use your senses to add descriptive details about the people and places in your memory.

Lesson 76—Use Examples: Writing Details

A. 1. For example, carbon dioxide is released into the atmosphere naturally from humans and plants. It is also released through unnatural sources, such as the burning of fossil fuels like coal, oil, and natural gas. Other sources of greenhouse gases include livestock, wetlands, and landfills, which release methane into the atmosphere.

2. One way is to talk to someone, like a parent or a teacher. … Another way to deal with stress is to take deep breaths.

B. Answers will vary. Write examples to support each of the main ideas provided.

C. Answers will vary. Write a short report about your favourite book, sport, or type of music. Use examples to support your main idea.

Lesson 77—Use Variety: Writing Details

A. 1. location   2. time   3. importance

B. Answers will vary. Choose one of the topics provided to write about. Then, decide how to arrange the details in your writing and write some details on the topic using the method you chose.

C. Answers will vary. Write a paragraph on the topic from Exercise B, using the details you arranged.

Lesson 78—Format a Speaker's Words: Writing Dialogue

A. 1. Peyton and Jonah were walking home from school. "Did you hear about that earthquake?" asked Peyton. "I wish there was something I could do to help."
"Yeah, me too," said Jonah. "Maybe we can hold a fundraiser to help the relief efforts."

B. Answers will vary. Finish writing the story, including dialogue that is formatted correctly.

C. Answers will vary. Write a story about being granted one wish. Include dialogue that is formatted correctly.

Lesson 79—Make Language Precise: Avoiding Redundancies

A. 1. ~~end~~   4. ~~ahead~~
   2. ~~exact~~   5. ~~the reason~~
   3. ~~together~~   6. ~~ahead~~

B. My dad and I led the way to the lake as my little brother followed us. When we got there, we looked across the frozen ice. I knelt to lace my skates, and then helped my brother with his. We walked onto the ice and started gliding slowly. The weather was perfect. In all of my experiences on the lake, I had never seen the ice so smooth.

C. Answers will vary.

Lesson 80 — Sum Up Your Narrative: Strong Conclusions

A. Answers will vary. Write strong conclusions that convey an emotion for the narratives described.

B. Answers will vary.

Lesson 81 — Sum Up Your Report: Strong Conclusions

A. Answers will vary. Write strong conclusions for the topics and main points provided. Your conclusions should leave readers thinking.

B. Answers will vary. Write a strong conclusion for the report.

C. Answers will vary. Write a report about a sport, including a strong conclusion that leaves readers thinking.

Lesson 82 — Catch Your Readers' Attention: Effective Titles

A. **1.** b **2.** a

B. Answers will vary. Rewrite the titles to make them catchier and format your revised titles correctly.

C. Answers will vary.

Lesson 83 — Check for Errors: Revising and Editing

A. 1. I thought the speech on sustainability was really interesting.
2. Ashton and Mara's plane arrived in Ottawa an hour late.
3. My sprained ankle got worse because I kept playing soccer after I injured it.
4. "I wish I had studied harder for the history test," said Farrah. "I don't think I did very well."
5. I ran as fast as I could, but I didn't catch up to my brother.
6. We learned about the diet, living conditions, and gender roles of people living in New France and British North America in the eighteenth century.
7. "Remember to study for the final geometry test," said Ms. Marble as we walked out of class.

B. Answers will vary. Rewrite the paragraph, using a variety of sentence lengths and correct capitalization and punctuation.

C. Answers will vary. Rewrite the paragraph, correcting any errors in capitalization and punctuation and using a variety of sentence lengths.

D. Answers will vary. Write a paragraph about what a house in the future might look like. Then, revise and edit the paragraph using the strategies from this lesson.

Lesson 84 — Correct Sentences: Revising and Editing

A. Answers will vary. Your rewrites should fix the sentence fragment *Like stepping back in time* and the run-on sentence *They were kept in a locked display case every now and then she would open it up and let me touch them*.

B. Answers will vary. Write several paragraphs about a childhood memory. Then, reread the paragraph and correct all the errors that you find.

Section Review

A. **1.** teacher / scientists   **2.** children

This exercise is a review for Lesson 69.

B. **1.** process chart   **3.** chronological order
2. chronological order

This exercise is a review for Lesson 72.

C. Answers will vary. Use dialogue to craft a strong opening for each of the story topics.

This exercise is a review for Lesson 73.

D. Answers will vary. Use your senses to write descriptive details about the characters and settings.

This exercise is a review for Lesson 75.

E. Answers will vary. Write at least two examples to support the main idea.

This exercise is a review for Lesson 76.

F. **1.** compare and contrast   **2.** time   **3.** location

This exercise is a review for Lesson 77.

G. "We're almost there," said Lewis as we trudged up what felt like an Everest-sized cliff. We were hauling all of our gear on our backs: tent, tarp, cooler, backpacks. I felt like my back was about to break into little pieces. Finally, we reached the top of the cliff.

"Look over there," he said, pointing off in the distance. "Do you see it?"

"Way over there? I thought you said we were close!" Just then, we heard a clap of thunder. I looked up. What had been a clear blue sky had turned dark and cloudy.

"Better keep moving," said Lewis, walking swiftly ahead of me.

This exercise is a review for Lesson 78.

H. 1. ~~free~~   2. ~~future~~

This exercise is a review for Lesson 79.

I. Answers will vary. Choose one of the topics and write a strong conclusion that leaves readers thinking.

This exercise is a review for Lessons 80 and 81.

J. 1. b   2. a

This exercise is a review for Lesson 82.

K. Answers will vary.

This exercise is a review for Lessons 83 and 84.

L. Answers will vary.

This exercise is a review for Lessons 73, 74, 76, 77, 81, and 82.

Develop Research Skills

Lesson 85—Have a Clear Focus: Inquiry Questions

A. 1. Ineffective. Explanations will vary. Note that the question is not open-ended, that it is not meaningful, or that the answer is not debatable.

2. Effective. Explanations will vary. Note that the question is open-ended, focused, and researchable.

B. Answers will vary. Write inquiry questions that meet the criteria listed in the box at the top of the lesson. Questions may or may not be debatable.

C. Answers will vary.

Lesson 86—Find Synonyms: Researching Words

A. 1. courage, fearlessness   3. dispute, altercation
2. thwart, interfere

B. Answers will vary. Possibilities include, but are not limited to, the following:

1. protected, secure   4. angry, enraged
2. cheerful, glad   5. tour, inspect
3. prompt, urge   6. mutter, mumble

C. 1. shouted   3. crumble   5. hated
2. appropriate   4. suspect

D. Answers will vary.

Lesson 87—Choose Resources: Library Research

A. 1. d   2. a

B. current election – newspaper; average annual temperature of Vancouver – almanac; maintain ATV engine – manual; pronunciation of a word – dictionary; book recommended by friend – library database

C. Answers will vary.

Lesson 88—Conduct Online Research: Keywords

A. Answers will vary. *Sample answers*:

1. WWF Canada president; World Wildlife Fund Canada president, date elected
2. Black Loyalist discrimination New Brunswick eighteenth century
3. Treaty Utrecht terms
4. overfishing Bay of Fundy; environment overfishing Bay of Fundy

B. Answers will vary.

Lesson 89—Know the Difference: Primary and Secondary Sources

A. 1. Primary: data from an experiment; diary; interview; autobiography; artifacts from an archaeological site

Secondary: encyclopedia; biography; book about the Seven Years' War; history book

B. Answers will vary. Benefits of secondary sources may include the following: more general coverage of the battle; combination of primary and secondary sources about the battle; possibly more balanced or complete analysis; may offer more than one perspective.

Benefits of primary sources may include the following: first-hand account; can see how someone going through the battle felt, gain of an understanding of what it was like to be there; more personal.

C. Answers will vary.

Lesson 90—Look for Consensus: Evaluating Websites

A. c

B. Answers will vary. *Sample answers*:

1. The website can be trusted. Government websites usually contain up-to-date, valid information, and the facts have been verified through other sources.

2. There is no way of knowing whether the author has valid statistics, so you should not use them. Unless you can find other sources that back up the statistics, you should not use them in your report.

Lesson 91—Be Critical: Evaluating Websites

A. Answers will vary.

1. Note that because the main purpose of the site is to sell merchandise, you should not immediately trust the information. However, if the information is supported by other sources, it may be considered reliable.

2. You should not trust or use the information since no sources of evidence are provided and you are not able to verify the claims made.

3. You should trust the information. Not all sites contain time stamps and, while they are helpful, all other evidence points to the information being reliable. The author is a specialist and the information is supported by other sources, including the bibliography.

B. Answers will vary.

Lesson 92—Remix and Rework: Plagiarism and Copyright

A. 2.

B. 3.

C. Answers will vary.

D. Answers will vary.

Lesson 93—Citations for Videos: Crediting Sources

A. 1. EarthMattersNow. "Clear-Cutting Rainforests in Madagascar." Online video clip. *YouTube*. YouTube, 12 Nov. 2011. Web. 26 Jun. 2015.

2. CineFocus Canada. "Green Heroes: The Art of Change." Online video clip. *Green Heroes TV*. TVO, 27 Dec. 2014. Web. 22 Sept. 2015.

B. Note that whenever you use information from researched sources, you must credit the creator in order to avoid plagiarism, follow copyright laws, and respect the work of others.

C. Answers will vary.

Lesson 94—Track Print and Online Sources: Research Notes

A. 1. c   2. a

B. Answers will vary. List books, magazines, websites, videos, photographs, music.

C. Answers will vary. Track online sources by bookmark number and print sources by author name, title, and page(s).

D. Answers will vary.

Lesson 95—Use Ideas and Words: Paraphrasing and Quoting

A. Answers will vary. Rewrite the information in your own words.

B. Answers will vary. Write a sentence that includes the quote and cites it correctly.

C. Answers will vary. Identify the main idea, choose an appropriate quote, and make point-form notes.

D. Answers will vary.

Section Review

A. 1. Ineffective. Answers will vary. Note that it is not open-ended or debatable and does not require much research to answer.

2. Effective. Answers will vary. Note that it is open-ended, debatable, and able to be answered through research.

This exercise is a review for Lesson 85.

B. hostile; strategy; hesitate; arrogant

This exercise is a review for Lesson 86.

C. Answers will vary. *Sample answers*: murky, filthy, cloudy

This exercise is a review for Lesson 86.

D. recent typhoon – newspaper; better words for an essay – thesaurus; repair a lawnmower – manual; book recommended by a friend – library database; topographical map of Canada – atlas

This exercise is a review for Lesson 87.

E. Answers will vary. *Sample answers*:
1. Alberta minister health
2. Canadian Pacific Railway impact development Canada

This exercise is a review for Lesson 88.

F. 1. P   2. S   3. P   4. P   5. P   6. S

This exercise is a review for Lesson 89.

G. b) ✓   d) ✓

This exercise is a review for Lesson 90.

H. Answers will vary.

This exercise is a review for Lesson 91.

I. False. Note that payment for material involving someone else's work cannot be accepted without the specific permission of the original creator.

This exercise is a review for Lesson 92.

J. Science For Tomorrow. "Producers, Consumers, and Decomposers." Online video clip. *YouTube*. YouTube, 22 Jan. 2013. Web. 3 Sept. 2015.

This exercise is a review for Lesson 93.

K. Answers will vary.

This exercise is a review for Lesson 95.

L. Answers will vary. Paraphrase the paragraph in Exercise K based on the notes you created.

This exercise is a review for Lesson 95.

M. Answers will vary.

This exercise is a review for Lesson 94.

N. Answers will vary.

This exercise is a review for Lesson 95.

STORY MAP

Characters	Setting

Problem	Solution

Beginning	Middle	End

WORD WEB

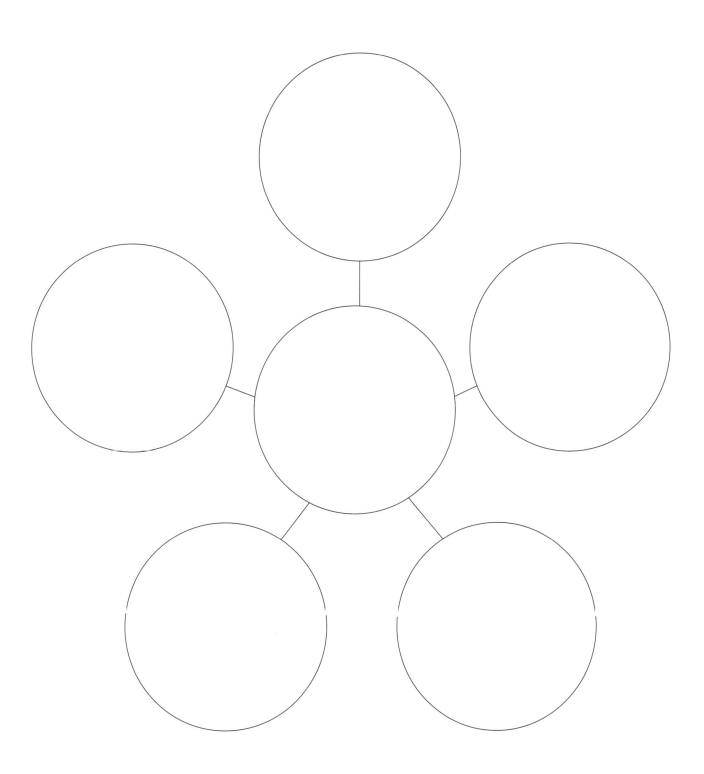

5W WEB

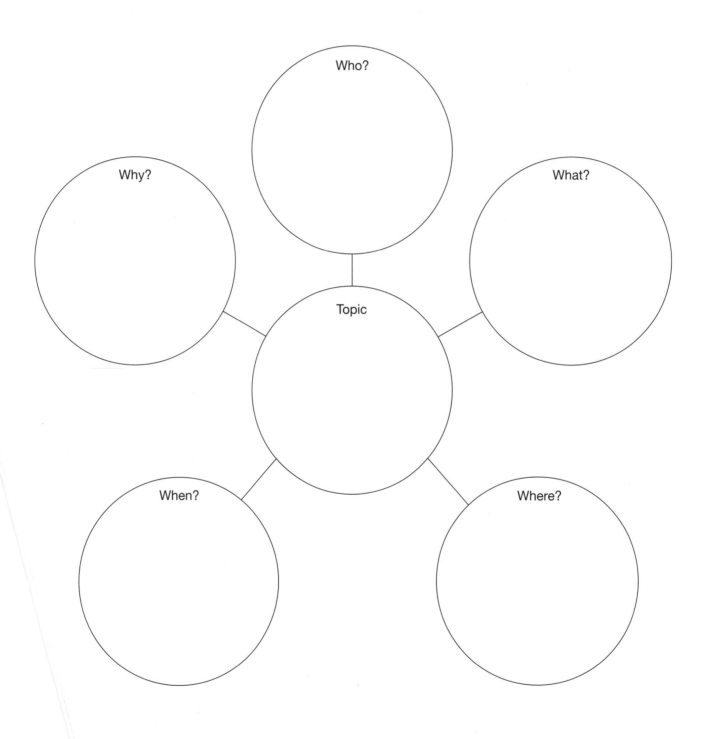

VENN DIAGRAM

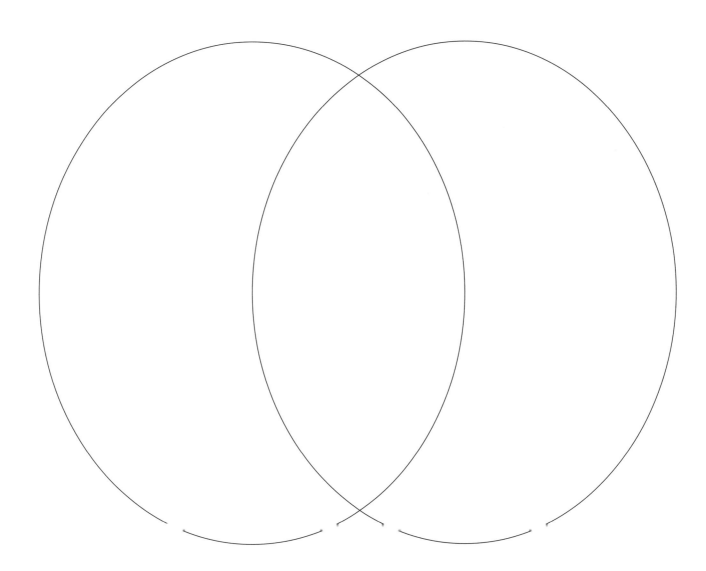

INDEX

Abbreviations
    appropriateness of, 54
    in dictionary, 53
    use of, 54
Adjectives
    adding details, 31–32
    adjective clause, 42
    adjective phrase, 31–32
    comparative form, 96
    definition of, 94, 99
    demonstrative adjectives, 94
    descriptive details, 119–120
    making comparisons, 96
    nouns as adjectives, 94
    positive form, 96
    possessive adjectives, 94
    prepositional phrases, 101
    superlative form, 96
    types of adjectives, 94–95
    use of, 94–95
    vs. adverbs, 99
Adverbs
    adding details, 31–32
    adverb clauses, 43
    adverb phrase, 31–32
    comparative form, 98
    definition of, 97, 99
    making comparisons, 98
    positive form, 98
    prepositional phrases, 101
    relative adverbs, 42
    superlative form, 98
    use of, 97
    vs. adjectives, 99
Alliteration, 19–20
Almanac, 142
Antecedent, 92, 93
Antonyms, 7
Apostrophes, 13, 59, 63, 72, 73, 74, 88
Base words, 10, 11, 12
Bibliography, 148, 149, 151
Capitalization, 51–52, 60, 61, 70, 131
Clauses
    adjective clauses, 42
    adverb clauses, 43
    definition of, 39, 60
    independent clauses, 39, 40, 55, 60
    main clause, 39, 40
    subordinate clauses, 39, 40, 43, 55, 105
Colons, 61, 63
Comma splice, 45
Commas
    and adverb clauses, 43
    in compound sentence, 29, 33
    and conjunctions, 29, 105
    in dialogue, 55, 125
    interjections, 106
    after participle phrases, 102
    after subordinate clause, 39, 40, 55
    use of, 55–56
Comparisons, 96, 98
Conjunctions
    and adverb clauses, 43
    compound sentences, 29
    coordinating conjunction, 45, 105
    definition of, 105
    fixing run-on sentences, 33
    and subordinate clauses, 39, 40
    subordinating conjunction, 44, 105
    in titles, 131
Connotation, 14
Contractions, 13, 54, 83
Copyright, 147
Denotation, 14
Details. *See* writing details
Dictionary, 14, 53, 142
Emotions, 106, 128
Encyclopedia, 142
Exclamation marks, 26, 106, 125
Gender-neutral language, 86–88
Homophones, 8–9
Idioms, 19–20
Interjections, 106
Internet
    online sources, 148, 149–150
    research, 143, 149
    videos, 148
Joining words. *See* conjunctions
Library database, 142

Literary devices, 19–20
Manual, 142
Metaphors, 19–20
Modifiers
    see also adjectives; adverbs
    dangling modifier, 104
    definition of, 103
    misplaced modifiers, 103, 104
Newspapers, 142
Nouns
    abstract nouns, 75
    as adjectives, 94
    antecedent, 92
    collective nouns, 70–71, 83
    common nouns, 70–71
    concrete nouns, 75
    definition of, 70
    irregular plural possessive nouns, 74
    plural nouns, 59, 73
    plural possessive nouns, 59, 73, 74
    possessive nouns, 72, 73, 74
    proper nouns, 70–71
    singular nouns, 59, 72
    singular possessive nouns, 59, 72
    specific nouns, 15
    strong nouns, 15–16
    types of nouns, 70–71
    vivid nouns, 15
Object
    definition of, 38
    direct objects, 38
    indirect objects, 38
    of the preposition, 101
Onomatopoeia, 19–20
Organizing ideas
    chronological order chart, 115–116
    clustering, 114
    graphic organizers, 115–116, 172–175
    life map, 111
    process chart, 115–116
    topic/subtopic web, 114
Ownership, 59, 72, 73, 74
Paraphrase, 151–152
Parentheses, 62, 63
Participle phrases, 102
Periods, 26, 55
Personification, 19–20
Plagiarism, 147
Possession. *See* ownership
Predicates
    complete predicates, 34–35
    simple predicates, 37
Prefixes, 10, 11
Prepositional phrases, 101
Prepositions, 100, 131
Pronouns
    agreement with antecedent, 92, 93
    definition of, 86, 87, 88
    gender-neutral pronouns, 86–88
    indefinite pronouns, 89
    object pronouns, 87, 90
    possessive pronouns, 88
    reflexive pronouns, 90, 91
    relative pronouns, 42
    subject pronouns, 86
    subject–verb agreement, 83
Punctuation marks, 26, 61, 62, 63–64, 125
    see also specific punctuation marks
Question marks, 26, 125
Quotation marks
    double quotation marks, 57–58, 63
    in research paper, 151
    single quotation marks, 57–58, 63
    use of, 57–58, 63, 125–126
Quotations, 151–152
Research process
    bibliography, 148
    citing research, 148, 149–150
    effective inquiry questions, 140
    keywords, 143
    library research, 142
    multimedia presentation, 149
    online research, 143, 149
    online sources, 149
    paraphrase, 151–152
    plagiarism, 147
    primary sources, 144
    print sources, 149, 151
    quotations, 151–152

    research notes, 149–150
    secondary sources, 144
    videos, citations for, 148
    websites, evaluating, 145, 146
    words, researching, 141
    works cited page, 149
Root words, 10
Semicolons, 45, 60, 63
Sentence fragments, 44, 134–135
Sentences
    complex sentences, 40–41
    compound sentences, 29–30, 55
    declarative sentences, 26–27
    editing sentences, 33, 44, 45
    exclamatory sentences, 26–27
    expand sentences, 31–32
    imperative sentences, 26–27
    independent clause, 39, 40
    interrogative sentences, 26–27
    length of sentences, 28
    revising and editing, 134–135
    run-on sentences, 33, 134–135
    sentence structure, checking, 134–135
    types of sentences, 26–27
Simile, 19–20
Speaker tags, 125–126
Statistics, 118
Subject–verb agreement, 83–84, 85, 89
Subjects
    agreement with verb, 83–84, 85, 89
    complete subjects, 34–35
    definition of, 36
    simple subjects, 36
Subtitles, 61
Suffixes, 10, 12
Synonyms, 6
Thesaurus, 141, 142
Titles, 61, 63, 131
Verbs
    action verbs, 76–77
    agreement with subject, 83–84, 85, 89
    auxiliary verbs, 76–77, 78
    future tense, 79–80
    irregular verbs, 81
    linking verbs, 76–77, 85
    participles, 82, 102
    past perfect tense, 82
    past progressive tense, 79–80
    past tense, 79–80, 81
    perfect tenses, 82
    present perfect tense, 82
    present progressive tense, 79–80
    present tense, 79–80, 102
    specific verbs, 15
    strong verbs, 15–16
    types of verbs, 76–77
    verb tenses, 79–80
    vivid verbs, 15
    wordy verbs, 15
Writing details
    arranging details, 123–124
    character, 119–120
    descriptive details, 119–120
    examples, 121–122
    setting, 119–120
    supporting details, 121–122
    variety, 123–124
Writing process
    audience, 17, 112, 113
    choosing a topic, 111
    dialogue, 57–58, 63, 117, 125–126
    editing, 33, 44, 45, 132–133
    expository essay, 113
    formal language, 17–18, 54, 112
    informal language, 17–18, 106, 112
    literary devices, 19–20
    narratives, 128
    purpose, 17, 112
    revising, 132–133, 134–135
    strong conclusions, 128, 129–130
    strong openings, 117, 118
    titles, 131
    topic *vs.* thesis, 113
    writing details. *See* writing details
Writing traits
    making language precise, 127
    organizing ideas, 111, 114, 115–116
    redundancies, avoiding, 127